T0185683

Octopuses, Squid & Cuttlefish

Ole G. Mouritsen · Klavs Styrbæk

Octopuses, Squid & Cuttlefish

Seafood for Today and for the Future

Ole G. Mouritsen
Department of Food Science
University of Copenhagen
Copenhagen, Denmark

Klavs Styrbæk
STYRBÆKS
Odense N, Denmark

Translation and adaptation to English by Mariela Johansen

Photography by Jonas Drotner Mouritsen

ISBN 978-3-030-58029-2   ISBN 978-3-030-58027-8   (eBook)
https://doi.org/10.1007/978-3-030-58027-8

© The Editor(s) (if applicable) and The Author(s), under exclusive license to Springer Nature
Switzerland AG 2018, 2021
This work is subject to copyright. All rights are reserved by the Publisher, whether the whole or part of
the material is concerned, specifically the rights of translation, reprinting, reuse of illustrations,
recitation, broadcasting, reproduction on microfilms or in any other physical way, and transmission or
information storage and retrieval, electronic adaptation, computer software, or by similar or dissimilar
methodology now known or hereafter developed.
The use of general descriptive names, registered names, trademarks, service marks, etc. in this
publication does not imply, even in the absence of a specific statement, that such names are exempt from
the relevant protective laws and regulations and therefore free for general use.
The publisher, the authors, and the editors are safe to assume that the advice and information in this
book are believed to be true and accurate at the date of publication. Neither the publisher nor the
authors or the editors give a warranty, expressed or implied, with respect to the material contained
herein or for any errors or omissions that may have been made. The publisher remains neutral with
regard to jurisdictional claims in published maps and institutional affiliations.

This Springer imprint is published by the registered company Springer Nature Switzerland AG
The registered company address is: Gewerbestrasse 11, 6330 Cham, Switzerland

Contents

List of Recipes

The People behind the Book

Ole G. Mouritsen

is a research scientist and professor of gastrophysics and culinary food innovation at Copenhagen University. His work focuses on basic sciences and their applications within the fields of biotechnology, biomedicine, and food. He is the recipient of numerous prizes for his work and for research communication. His extensive list of publications includes a number of monographs, several of them co-authored with Klavs Styrbæk, which integrate scientific insights with culinary perspectives and have been nominated three times for Gourmand Best in the World Awards. Currently, Ole is president of The Danish Gastronomical Academy and director of the National Danish Taste Centre *Taste for Life*, which is generously supported by the Nordea Foundation. This is a cross-disciplinary centre that aims to foster a better understanding of the fundamental nature of taste impressions and how we can use this knowledge to make much more informed and healthier food choices. Its extensive educational program reaches out to audiences of all ages, with a special effort directed toward children to shape their dietary habits from an early age. For many years, Ole has been fascinated with the Japanese culinary arts and in explaining the extent to which its techniques and taste elements can be adapted for the Western kitchen. In recognition of his efforts, he was appointed in 2016 as a Japanese Cuisine Goodwill Ambassador by the Japanese Ministry of Agriculture, Forestry, and Fisheries, and in 2017 the Japanese Emperor bestowed upon him The Order of the Rising Sun, Gold Rays With Neck Ribbon, *Kyokujitsu chujusho* 旭日中綬.

Klavs Styrbæk

is a professional chef who owns and operates STYRBÆKS together with his wife, Pia. By combining a high standard of craftsmanship, sparked by curiosity-driven enthusiasm, he has created a gourmet centre where people can enjoy excellent food and where they can come to learn and take their culinary skills to a whole new level. Klavs is particularly enthusiastic about seeking out unique, local raw ingredients that are incorporated into new taste adventures or used to revisit traditional Danish recipes that might otherwise be forgotten. This delicate balance between innovation and renewal is demonstrated in his award-winning cookbook *Mormors mad* (*Grandmother's Food*) (2006), which was honored with a special jury prize at the Gourmand World Cookbook Awards in 2007. In 2008 and 2019 he was awarded an honorary diploma for excellence in the culinary arts by the Danish Gastronomical Academy. Many of the recipes that appear in the books co-authored with Ole originated in the test-kitchens at STYRBÆKS.

Jonas Drotner Mouritsen

is a graphic designer and owner of the design company Chromascope, which specializes in graphic design, animation, and film production. His movie projects have won several international awards. In addition, he has been responsible for layout, photography, and design of several books about food, some of which have been nominated for Gourmand World Cookbook Awards.

Mariela Johansen

who has Danish roots, lives in Vancouver, Canada, and holds an MA in Humanities with a special interest in the ancient world. Working with Ole and Klavs, she has translated several monographs, adapting them for a wider English language readership. Two of these, *Umami: Unlocking the Secrets of the Fifth Taste* and *Mouthfeel: How Texture Makes Taste*, won a Gourmand World Cookbook Award for the best translation of a cookbook published in the USA in 2014 and 2017, respectively.

Acknowledgements

In our work on this project we have been helped and guided by a long list of colleagues and good contacts. We are particularly indebted to:

- The members of the 'The Squid Squad' at Nordic Food Lab – Louise Beck Brønnum, Charlotte Vinther Schmidt, Yi-Ting Sun, Peter Lionet Faxholm, Roberto Flore, and Karsten Olsen. Special thanks are due to: Karsten Olsen, who enthusiastically and imaginatively engaged himself in gatrophysical and gastronomic studies of Danish squid; Roberto Flore, who arranged a field trip to Sardinia to see how cephalopods are caught and to sample how they are used in local gastronomic specialties; Peter Lionet Faxholm, who sorted out taxonomic relationships, reviewed the manuscript thoroughly, and made valuable comments on it.
- Co-workers at *Taste for Life* for inspiration and joint work on cephalopod taste.
- Mathias Porsmose Clausen, University of Southern Denmark, for microscopy of cephalopod muscle structure.
- Prof. Anna Di Cosmo and Prof. Gianluca Polese, Università degli Studi di Napoli Federico II, for discussions about octopuses and their nervous system, as well as the necessity to exploit marine resources more sustainably.
- Dr. Graziano Fiorito, Stazione Zoologica Anton Dohrn, Naples, for informative discussions about Cephalopoda, especially concerning the behavioural biology of octopuses and the significance of implementing rules for the treatment of cephalopods, and also for the present of his impressive book on pattern formation in cephalopods.
- Dr. Pamela Imperadore, Stazione Zoologica Anton Dohrn, Naples, for a tour of the laboratories with the octopus aquaria, as well as insightful explanations about the animals' behaviour, their nervous systems, and their ability to regenerate them if damaged.
- Dr. Xavier Bailly, Station Biologique de Roscoff in Brittany, for a tour of the institution and an explanation of his laboratory model for aquaculture of *Sepia officinalis*, which has enabled him to raise cuttlefish through a complete life cycle, from eggs to adults.
- Prof. José Lucas Pérez Lloréns, Cádiz, for information about Andalusian *pescaíto frito* made with cephalopod meat and for arranging contacts with Spanish cephalopod researchers.
- Dr. Ling Miao for information about the Chinese expression for deep-fried squid.
- Captain Ignacío López Cabrera (Cofradía de Pescadores de Motril), Abdelaziz Shabi (Grupo de Pesca Costa Granada), and biologist Inmaculada (Inma) Carrasco Rosada for arranging a fishing trip on a trawler along the coast near Motril and Almuñécar in Southern Spain. Special thanks to Inma for information about species designations for cephalopods and fish.

- Head chef Yoshinori Ishii (Restaurant Umu, London) and Sakiko Nishihara for conversations about Japanese ways of preparing cephalopods.
- Poul Rasmussen, Albanifisk, for supplying fresh cephalopods from the North Sea.
- Dr. Paul John Frandsen for carefully identifying the images of cephalopods in the Punt reliefs.
- Prof. Masayoshi Ishida, Ritsumeikan University, Japan, for the outstanding gift of *surume*, traditional dried Japanese flying squid.
- Mette Holm for references to cephalopods in Japanese cuisine.
- Jens Peter Henriksen for bringing back fresh cephalopods from Malaga in this hand luggage.
- Rikke Højer for making an imprint, *gyotaku*, using a Danish squid.
- Prof. Paw Dalgaard for information and literature concerning the risks associated with eating cephalopods.
- Mads Friis Nielsen and Restaurant mmoks for lending serving plates for food photography.
- Steen Aalund Olsen for the loan of a belemnite.
- Kristoff Styrbæk for several of the pictures of food included in the book.
- Chef Kasper Styrbæk for participating in the development of cephalopod recipes and dishes.

This book first took shape when one of us (Ole) had a grant that allowed him to spend some productive writing time in May 2017 at the thousand-year-old monastery of the San Cataldo Institution. Its peaceful atmosphere, together with its surroundings on the Amalfi coast of southern Italy, a UNESCO World Heritage site of immense natural beauty and many architectural and art treasures, provided the perfect framework for delving into the strange and fascinating world of the cephalopods and, especially, their culinary uses. At this point Klavs had already, over a period of years, experimented with preparing cephalopods in his professional kitchen at STYRBÆKS.

In addition, this book draws on a wide range of scholarly articles and scientific literature. Many individuals and organizations have generously made photographs, images, and other illustrative materials available for use in this book. See Bibliography and Illustration Credits.

Jonas Drotner Mouritsen has participated in the production of this book from its inception and has taken many of the photographs.

This book was originally written and published in Danish, the mother tongue of the authors. The present volume is a fully updated and carefully revised version that was translated and adapted for a broader international audience by Mariela Johansen. Mariela undertook the challenging task of working with the interdisciplinary material to produce a coherent, scientifically sound, and very accessible book. This involved not only translating the text, but also checking facts, ensuring consistency, and suggesting new material and valuable revisions. The authors are extremely appreciative of her devotion to this project.

We would also like to thank our editor, Daniel Falatko, for his enthusiastic response to the book and interest in seeing it through to publication, and Springer Nature for the professional and expeditious handling of the manuscript.

Last, but not least, we wish to extend our heartfelt thanks to our families, especially Kirsten and Pia, who have been at our side as we pursued our journey through the world of cephalopods in the course of a couple of years. They have patiently lent us their ears and, on numerous occasions, their palates, and have been a constant source of love and support.

Introduction

Octopuses, squid, and cuttlefish are some of the strangest creatures inhabiting this planet, living in oceans in all parts of the world both in shallow waters and many kilometres under the surface. Throughout the ages their mysterious presence in the deep dark seas has both fascinated and frightened humans, giving rise to legends, myths, fantasies, and stories that have found their way into art, literature, and film. Much about them seems foreign and exotic—their unusual appearance with many appendages attached to their head, frightening looking suckers, soft, pliable bodies with an uncanny ability to alter their shape and change colour, and an acrobatic way of moving around, sometimes at lightning speed.

Together with their cousins, the hard-shelled nautiluses, octopuses, squid, and cuttlefish are the sole surviving members of a remarkable group comprised of about 800 species of marine animals that are classified as cephalopods. They are the descendants of an ancient life form that in the course of 500 million years has evolved in the oceans as part of the biological phylum of invertebrates called molluscs (Mollusca). Even though it is hard to understand based on their appearance alone, they are closely related to bivalves such as oysters, mussels, and clams and to gastropods such as snails and slugs.

The name *cephalopod* is derived from the Greek words *kephalē* (head) and *pous* (foot), a highly descriptive indication of the way in which the limbs are attached to the main body of the animal. Modern biologists have actually made things somewhat more complicated by adopting a different set of terms for cephalopod anatomy, so it is useful to clear this up at the outset. Octopuses, squid, and cuttlefish all have eight appendages that are referred to as *arms*, but squid and cuttlefish generally have an additional two that are called *tentacles*. The bulky part (body) of all three types is usually divided into the *head* and the *mantle* to which *fins*, also referred to as *wings*, are attached on many of the species. In some other languages and among fishermen the soft-bodied cephalopods are known as 'inkfish' and 'ink squirters,' a reference to the dark coloured liquid some species use as a defence mechanism.

Some species of the soft-bodied cephalopods can grow to an enormous size, up to eighteen metres in length and with eyes as large as soccer balls, yet they live for no more than a couple of years. Others are small and delicate. They mate only once in the course of their lifetime and die shortly after they have reproduced. They have blue blood and three hearts and can eject jet black ink. They have a brain and exhibit behaviours that could be called intelligent but, oddly, most of their central nervous system is located in the arms. The colour of their skin and the patterns on it can change in a split second, but they are actually colour blind. They have taste sensors in their arms. They are masters of disguise, able to take on the appearance of rocks and gravel or of other sea creatures. They can propel themselves to swim with lightning speed and shoot out a tentacle to catch prey at the speed of a javelin thrown by an elite athlete. On the other hand, the hard-shelled nautiluses, which can live for up to 15 or 20 years, have changed so little since they first appeared on evolution charts that they are considered to be living fossils. Is it any wonder that from time immemorial cephalopods have fascinated people all over the world?

While the nautiluses are valued primarily for their beautiful shells, octopuses, squid, and cuttlefish are soft-bodied and highly edible. But they have been largely overlooked in the culinary traditions of many countries, where people have mostly come across them as breaded, deep-fried, and often rubbery squid rings, in ready-made seafood salads, or as small pieces of octopus arm found on a platter of assorted sushi. In some parts of the world, however, they are highly regarded and have for many centuries been prepared in traditional ways to create delicious dishes. They come in such great variety that they present a wide range of options for making healthy and nutritious meals, especially given that they are protein-rich and have little fat. And they should be appreciated not only for their taste but also for their special texture and mouthfeel. While their mantles and arms are the most prominent parts of these sea creatures, their ink, mouth parts, and innards are also used in some dishes. They can be eaten in so many ways—raw, cooked, grilled, marinated, dried, smoked, and fermented—the possibilities are almost endless.

Where cephalopods are not commonly eaten, they often show up as incidental bycatch associated with fisheries for other species and are regarded as a nuisance to be sold off as a low-value source of bait. This might change, however, as the total world catch of marine and freshwater finfish may already be at the maximum level that is sustainable, leading to a shift toward more 'unconventional' food sources. And this is where octopus, squid, and cuttlefish come into the picture. Recent studies have shown that they appear to be thriving, with their rapid population growth an indication that they may even be benefitting from a changing ocean environment. Given that squid and octopuses are known to engage in cannibalism and that the stocks of some species are increasing, we should eat them before they eat each other! They have the potential to become a more important protein source to supplement our consumption of meat from land-based animals. By utilizing them as food we can also find new ways to make use of the oceans' resources in a more insightful and responsible way. So there are many good reasons to look forward to seeing them much more frequently on our dinner plates.

In this book we have set out to introduce our readers to all aspects of these fantastic creatures. After a brief look at some of the legends and myths that have come down to us through the ages, we move on to present some background information about cephalopods. What exactly are they, where do they come from, how do they live, what are their special characteristics, and how and where are they caught? Next we provide an overview of their taste and nutritional qualities and of the various techniques for preparing them for use as raw ingredients. This will lead us into the kitchen to explore the many possibilities for cooking with octopuses, squid, and cuttlefish—arms, heads, innards, ink, and all. We hope that the information and recipes in this section will enable just about anyone to turn out both everyday meals and gastronomic specialties that can be enjoyed as a delicious, exciting change from the usual. The book concludes with a discussion of how cephalopods might come to feature more prominently in, and make up a greater proportion of, the food that we will be eating in the future. Finally, to assist the reader we have included a reference section with a list of the various species discussed in the book and a glossary of technical and culinary expressions.

Strange Beings from the Depths of the Sea

» *"Since the sea is infinite and of unmeasured depth, many things are hidden …"*
Oppian of Anazarbus, *Halieutica*
(second century Greco-Roman poet)

Evidence of the importance attached to cephalopods can be found in the extent to which they feature in the visual arts of many cultures from ancient times right up to the present. Octopuses were particularly popular, being depicted on vases, pitchers, and other everyday objects and found in more artistic settings such as mosaics and frescoes. They appear as symbols on coins, jewelry, and ornaments. These sea creatures also caught the eye of poets, philosophers, and historians, who described them extensively both in classical writings and in modern works.

◘ Vase with an octopus motif from the Minoan culture on Crete, dating from ca. 1500 BCE.

In his great opus on natural history, *Historia Animalium*, written in the fourth century BCE, Aristotle sets out his observations about cephalopods referring to them as *octopus*, *sepia*, and *calamari*, which we might recognize as our modern-day octopuses, cuttlefish, and squid. He states that the *teuthus*, his name for a giant squid, can grow to a length of ten 'ell,' equivalent to about ten arms-lengths or four and a half metres. The Roman historian, Pliny the Elder (23–79 CE), also tackled the subject in his comprehensive work, *Naturalis Historia*, recounting the rather incredible story of a giant sea creature with a head "as big as a cask," arms that were about nine metres long, and weighing 320 kilograms. This squid-like animal was able to come onto land and steal fish that were being salted in open pickling tubs on the coast of Spain.

In the largest and most beautiful temple of ancient Egypt, Deir el-Bahari, dating from ca. 1465 BCE, located on the west bank of the Nile opposite Luxor, there is a relief that shows a squid as part of a row of marine animals.

Mythological Sea Monsters

Descriptions of truly frightening monsters and otherworldly sea creatures that terrify and threaten the lives of sailors can be found in mythologies and stories from many parts of the world. That such animals should loom large in the imagination is perhaps quite understandable given the dangers of going to sea in the small and relatively fragile boats of an earlier era.

In Homer's *Odyssey*, composed toward the beginning of the seventh century BCE, there is an account of Odysseus's dangerous passage through the Strait of Messina between Sicily and Calabria, where he must navigate between rocks inhabited by frightening creatures known as Scylla and Charybdis. The many-tentacled Scylla has become synonymous with a giant sea monster, which was later portrayed as an animal that resembles an octopus.

The Icelandic saga of Örvar-Oddr tells of a journey Oddr made with his son Vignir across the Greenland Sea to Helluland (Baffin Island). In the course of their voyage, they encounter huge creatures called *Hafgufa* (sea mist) and *Lyngbakr* (heather hill). The latter is described as the largest whale in the world. *Hafgufa*, on the other hand, is thought to be a veritable leviathan; it can swallow boats, people, and whales and stays submerged for days on end.

Hafgufa has been interpreted as a reference to the horrifying mythological giant sea monster *Kraken*, which might be either a squid-like or an octopus-like creature. *Kraken* is

"The octopus is a stupid creature, for it will approach a man's hand if it be lowered in the water; but it is neat and thrifty in its habits: that is, it lays up stores in its nest, and, after eating up all that is edible, it ejects the shells and sheaths of crabs and shellfish, and the skeletons of little fishes."

Aristotle, *The History of Animals*

a Norwegian word that has connotations of a being that is frightening and has a twisted appearance. In the course of the past century and a half, however, the legendary aspect of the *Kraken* has begun to unravel. Based on eyewitness accounts and the discovery of some enormous specimens that have washed ashore, it has been possible to conclude that the legend was probably based on sightings of giant squid of the species *Architeuthis* and not merely the stuff of overactive imaginations.

◘ Sketch of a giant squid that washed ashore in Fortune Bay, Newfoundland, in 1871.

With rum being associated so closely with sailors, it is appropriate that a popular Caribbean rum should be called Kraken and feature a picture of the dreaded sea monster on its label. Its rich brownish-black colour calls to mind the dark ink squirted by its namesake.

Theophilus Piccot gave an account of a dramatic encounter in Conception Bay, Newfoundland on the 26th of October, 1873, while fishing for herring in a small rowboat together with another fisherman. They suddenly spied a large and mysterious object floating on the surface of the water. In the hope that it might be something of value from a shipwreck, they rowed over to the object and saw that it was a large, quivering mass. When they poked it with a gaff it turned out to be alive and possessed of a beak "as big as a six-gallon keg." The creature rammed the bottom of the boat with the beak and two of its large arms started to hurl themselves around the boat in an effort to sink it. Using an axe, Piccot hacked away at its arms, which were as thick as a man's wrist. Ejecting a cloud of black ink into the water, the animal disappeared and was not seen again. Piccot brought the two severed arms back to shore, where one was put on display in the museum in St. John's. It is possible that this was an arm of the giant squid *Architeuthis*, although this is rather doubtful as we now know that these creatures live at great depths in the ocean and none have been detected alive near the surface of the water.

Over time whalers and other seafarers have reported sightings of large marine animals that are not whales, but in the majority of cases it is hard to document their observations and distinguish between the real and the imaginary. The accuracy of the details regarding the incident described by Theophilus Piccot in 1873 may be somewhat open to question. But despite our skepticism, when taken together with the subsequent scientific description of his discovery by more learned individuals in the community, this may well have been the first tangible evidence of the existence of the giant squid.

Was the Sea Monk of 1546 a Giant Squid?

During the reign of the Danish King Christian III, a peculiar marine animal was caught in the waters of Øresund, the strait between Sweden and Denmark. It caused a sensation all over Europe. The animal was described as being as big as a man, with a clean-shaven head, a human-like face, two large fins, and wearing something resembling a monk's habit that was covered with scales. On account of its bald pate, which looked like a tonsure, it was given the name 'sea monk.' This amazing discovery was reported to the king who had it brought to the castle and placed in the moat, where it died after a few days.

Three centuries later the Danish naturalist, Japetus Steenstrup (1813–1897), shed some light on the identity of this mysterious creature. When he had a chance to examine a similar specimen that had washed ashore in northern Jutland, he concluded that the 'sea monk' was actually a giant squid, giving it the name *Architeuthis monachus*. More recent interpretations have suggested that the creature from Øresund was most probably an angel shark (*Squatina squatina*). Nevertheless, as the remains of the original 'sea monk' were buried in an unknown location, its true nature will remain an enigma.

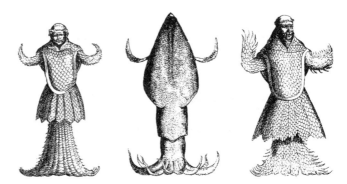

◧ Illustration of a curious marine animal, which was given the name 'sea monk,' that was found in the strait between Sweden and Denmark in 1546. At one time it was thought to be a giant squid.

◧ Antique print showing a giant octopus attacking a sailing ship (Pierre Denys de Montfort, 1766–1820). Natural History Museum, London.

We now know that there really are enormous octopuses weighing up to 200 kilograms and giant squid that weigh up to 500 kilograms with a body two to three metres long and an overall length of twelve to fifteen metres. Nevertheless, we have not yet learned much about how they live because they dwell in very deep waters. Until recently our knowledge about *Architeuthis* has been based on between 600 and 700 pieces that washed ashore from squid that had died, as well as the stomach contents of sperm whales, which prey on giant squid.

The first observations of a live giant squid were made in 2005 when a video camera was lowered to a depth of 900 metres in the north Pacific. The creature was lured into range with bait and a piece of tentacle was caught and brought to the surface, making it possible to identify the species and determine that the animal was at least eight metres in length. Later, in 2008, a team of researchers in a submersible with a specially built camera were able to encounter a four-metre long giant squid 'face-to-face' at a depth of 700 metres in the ocean south of Japan. The expedition resulted in actual footage of the animal, which showed it to be very active, a strong swimmer, and a very aggressive predator. In 2019 an expedition mounted by the National Oceanographic & Atmospheric Association succeeded in filming a giant squid in its natural habitat at a depth of 759 metres in the Gulf of Mexico. The video showed a juvenile specimen measuring about three to three and a half metres.

The giant squid Architeuthis was first identified as a separate species in 1857 by the Danish naturalist Japetus Steenstrup (1813–1897). Since then more than twenty specimens of Architeuthis have been described and until recently it was thought that there were several different species. Subsequent genetic research has shown, however, that all known examples of giant squid belong to the same species, *Architeuthis dux.*

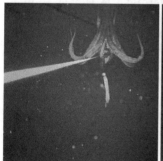

◘ First pictures of a living example of a giant squid, *Architeuthis dux*, taken in 2005 and 2012, respectively.

Ambergris (from Old French for 'grey amber') is a special secretion that is formed in the intestines of the sperm whale in the shape of grey rock-like lumps. These normally weigh between fifteen grams and fifty kilograms, but can be much heavier. The largest one ever was found in Antarctica in 1953 and weighed 420 kilograms. The lumps often contain the remains of giant squid, especially their beaks, and are usually either vomited or expelled as feces. In earlier times ambergris was a very valuable component in perfumes and aphrodisiacs, and was used to flavour food and taken as a medicine.

Researchers have estimated that there are now approximately 200 million giant squid world-wide. This number, which is at best a guesstimate, is derived from a rough calculation of the global sperm whale population (ca. 360,000) together with an assumption of how many giant squid a sperm whale eats, based on the number of their beaks found in its stomach contents.

Even though we have recently gained much more factual information about giant squid, there is no documented evidence that they have ever attacked or bitten humans, to say nothing of dragging a boat beneath the waves.

Cephalopods in Art, Literature, and Iconography

Over thousands of years, cephalopods have fascinated and inspired artists, writers, and filmmakers. The octopus, with its eight arms covered with rows of suckers and its strange behaviour, has not only frightened and conjured up dread in humans, but has also held a mystical and unnatural attraction for them, often with sensual and erotic overtones. In artistic works it is not unusual to find a careful balance between shivers of fear, pain, and death on one side and enjoyment, pleasure, and lust on the other. Should one be afraid or not? The latter probably finds its most graphic expression in the famous woodblock print, *Girl Diver and Two Octopuses* (1814), by the renowned Japanese artist Hokusai. The picture, which is a prime example of *shunga* (erotic art), is open to interpretation, signaling both pain and pleasure. It circulated widely in Europe in the second half of the eighteenth century, serving as inspiration for numerous writers and artists. Victor Hugo and Émile Zola incorporated the theme into their novels and painters such as Henri de Toulouse-Lautrec and, later, Pablo Picasso imported the motif directly into their own works. Likewise, octopuses and giant squid have served as models for visual artists and sculptors.

◘ The most famous example of a *shunga* wood-block prints, *Girl Diver and Two Octopuses* (1814) also known as *The Dream of the Fisherman's Wife*, by the Japanese master Katsushika Hokusai (1760–1849). The picture shows a sexual encounter between two octopuses and a woman.

Through the ages enormous cephalopods have also made their way into popular fiction, the best-known of which is probably Jules Verne's book, *Twenty Thousand Leagues under the Sea* (1869). It describes Captain Nemo's submarine, the *Nautilus* (eponymously named after the hard-shelled sea creature!), which is equipped with a glass window through which he and the others on board are able to observe an enormous octopus. A pack of these fearsome sea creatures later attack Nemo's vessel, tearing open the hatch and forcing the sailors to do battle against them with axes. In other fictional accounts, octopuses make an appearance as aggressive, fright-inducing colossal monsters that lurk at the bottom of the sea and in sunken ships guarding valuable treasures.

◘ Original illustration by Alphonse de Neuvilles of an octopus appearing outside Captain Nemo's submarine in Jules Verne's *Twenty Thousand Leagues under the Sea* (1869).

Both as a concept and in concrete form, octopuses have played a role in many comic strips and animated films, as well as horror and action movies. The best-known of these may be the James Bond thriller *Octopussy* (1983), which sets the erotic in sharp contrast to the gruesome. It features an octopus cult made up of beautiful women who smuggle jewels. Their leader is a business woman whose father kept, as a pet, a blue-ringed octopus, one of the world's most venomous marine animals. The combination of beautiful women, jewelry, and an octopus may have been inspired by the Japanese legend *Taishokan*, dating from the seventh century. It relates the struggle of a girl diver to recover a precious jewel stolen by the king of the sea dragon-serpents, sometimes pictured as an octopus.

"One has the sense of intelligent behaviour, of a regard that is much more expressive than that of some fish, to say nothing of marine mammals."

Jacques Cousteau (1973)

Another example is drawn from a more recent motion picture series, *Pirates of the Caribbean*, in which Captain Davy Jones of the ghost ship *The Flying Dutchman* is portrayed as having a face made up of an undulating mass of octopus-like arms with suckers, which twist and turn when he speaks or moves. His appearance can also be seen as a reference to the hideous female gorgons of Greek mythology. The most famous of these is Medusa, whose hair has been replaced by writhing, poisonous snakes.

And They Continue to Fascinate Us

Cephalopods are an ancient form of life, which over the passage of millions of years have shown themselves to be enormously successful. The more we find out about them and their peculiar existence, the more they mystify us and continue to captivate our imaginations. The numerous characteristics that set them apart—the large number of species, their varied shapes and sizes, their many long arms, their amazing ability to squeeze through the smallest spaces, their skill in going after prey and in escaping their predators, their well-developed nervous system and natural intelligence, their three hearts, the chameleon-like way that they can change colour and appearance in the blink of an eye—all contribute to making them something completely special and enigmatic.

We can add the focus of this book to the list, namely that these marine creatures are an overlooked, but versatile and tasty, food resource that can be used in so many ways to prepare a whole range of dishes that take advantage of their particular properties as raw ingredients, not least with regard to their contribution to creating a more exciting mouthfeel.

"Then the creeping murderer, the octopus, steals out, slowly softly, moving like a gray mist, pretending now to be a bit of weed, now a rock, now a lump of decaying meat while its evil goat eyes watch coldly. It oozes and flows toward a feeding crab, and as it comes close its yellow eyes burn and its body turns rosy with the pulsing colour of anticipation and rage. Then suddenly it runs lightly on the tips of its arms, as ferociously as a charging cat. It leaps savagely on the crab, there is a puff of black fluid, and the struggling mass is obscured in the sepia cloud while the octopus murders the crab."

John Steinbeck, *Cannery Row* (1945)

◼ North Pacific giant octopus (*Enteroctopus dolfleini*).

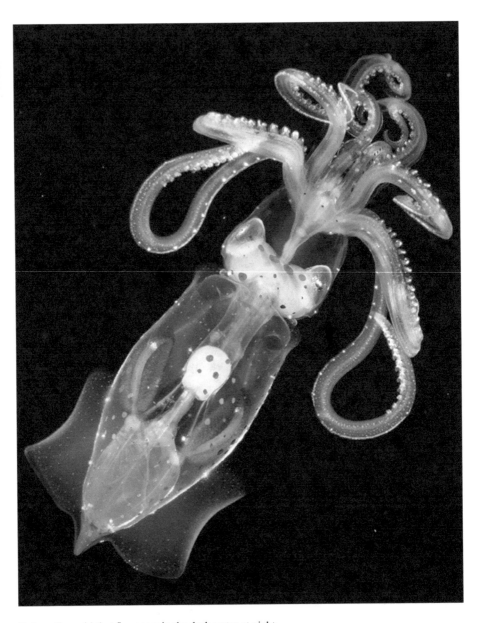

■ Juvenile squid that fluoresces in the dark water at night.

Such Abundance, So Much Diversity

Let us begin by looking at how the Cephalopoda, to use the scientific term, fit into the animal kingdom. They make up a class belonging to the phylum Mollusca (popularly called molluscs), which is the second largest classification of invertebrates and is made up of at least 100,000 known species. Most probably there are an equal number that have not yet been identified. The word mollusc is derived from the Latin *mollis*, which means 'soft,' an indication that these creatures are lacking a backbone.

It can often be difficult to give a definitive measurement of the size of a cephalopod because it depends on the extent to which the arms, and especially the tentacles, have been stretched. This is especially true of the squid and cuttlefish, where the tentacles are sometimes retracted right back to the mouth and at other times extended to reach a length that is many times that of the mantle. What is clear, however, is that cephalopods come in such an incredible range of sizes, from the tiny to the gigantic, that it staggers the imagination. These extremes can best be illustrated by a comparison between the heaviest adult squid specimen on record and that of the lightest. It has a mass that is about two and a half million times greater.

Some octopuses can be dangerous! The most poisonous are members of the four species of *Hapalochlaena* spp., known as the blue-ringed octopuses, which are found in shallow waters of the Pacific from Australia to Japan. Their salivary glands secrete a venomous neurotoxin, tetrodoxin, that is released when they bite their prey. It is so potent that it leads to paralysis and can be fatal even in minute quantities. Luckily they rarely attack unless provoked.

The giant and colossal squid are among the largest organisms on earth. The giant squid, *Architeuthis*, can grow to a length of fifteen metres, have a mantle that is about two metres long, have huge eyes, and can weigh hundreds of kilograms. The largest of all the cephalopods are the colossal squid, *Mesonychoteuthis hamiltoni*, that live in the freezing waters surrounding Antarctica. It is estimated that they might grow to a length of about eighteen metres, with a mantle that is three metres long, and weigh up to about 900 kilograms. Unfortunately our knowledge of them is severely limited as they live in remote parts of the ocean at depths of about a kilometre and to date only a few specimens have been caught. By way of contrast, the tiny squid *Idiosepius thailandicus*, which as their name suggests live in the Indian and Pacific Oceans near Thailand, grow to a length of only 0.1 centimetre!

Octopuses are of much more modest size and generally do not grow to be nearly as large as their squid cousins. The largest of them is the North Pacific species called *Enteroctopus dofleini*, which are thought to have an unusually long life span of up to five years. Normally octopuses attain a length of about five metres and can weigh up to fifty kilograms. The biggest one that has been caught to date measured 9 metres and weighed 272 kilograms. Here again there are dwarf species. The smallest are the *Octopus wolfi* that measure only two and a half centimetres and weigh less than one gram.

A Few Words about Their Classification

In this volume we are mainly concerned with those cephalopods that are readily available and that have gastronomic potential. These belong to the orders that are formally known as Octopoda (octopuses), Sepioidea (cuttlefish), and Oegopsida and Myopsida (squid). As already mentioned, octopuses have eight arms, while cuttlefish and squid have eight arms and two tentacles.

At present there are about 800 known species of cephalopods and there is probably an equal number that have not yet been identified. These are distributed approximately as follows: Octopoda – 300 species (of which 100 or so are members of the *Octopus* genus); Oegopsida and Myopsida – 300 species; Sepioidea – 120 species. The remaining species are the hard-shelled nautiluses (Nautilida) and some other minor groups of cephalopods.

Although we are sure that you will probably not need to know the actual scientific names of the different species when you set out to fish for cephalopods or to buy them to make dinner, this information might be useful as they are referred to later in the book and in some of the recipes. The ones most commonly eaten include the octopuses *Octopus vulgaris* and *Eledone cirrhosa*, the cuttlefish *Sepia officinalis*, and the squid *Loligo forbesii*, *Loligo vulgaris*, *Alloteuthis subulata*, and *Todarodes pacificus*. These and others, together with their common names in English, are listed in the supplementary information section at the back of the book.

Teuthida is an older designation for an order that included all the squid. Based on a closer study of their anatomy, squid have been reclassified as belonging to one or the other of two separate orders, Oegopsida and Myopsida.

'Head-Feet' that Have No Feet

As we have already discovered, the term cephalopod derived from the Greek words for head (*kephalē*) and foot (*pous*) is a misnomer. While their appendages correspond to what is called a foot on other molluscs, such as some bivalves and snails, they do not walk on them in the conventional sense, and the correct way to refer to them is arms.

■ Representative examples of three types of cephalopods: Octopoda (octopus), Sepioidea (cuttlefish), and Myopsida (squid).

All cephalopods have appendages. Those with eight arms only belong to the sub-order Octopodiformes (octopods) and those that have eight arms and two tentacles are in the sub-order Decapodiformes (decapods). On octopuses and on some species of other cephalopods one of the arms on the males (the hectocotylus) has been modified to be able to transfer sperm to the female during reproduction. The two tentacles on the decapods, which are located between the third and fourth pair of arms, are very long and used to capture prey. Some species are able to retract the tentacles into a pocket at their base from which they can rapidly be shot out. The other family of cephalopods, the Nautilida (nautiluses), have many more appendages, called cirri, about fifty in the case of females and ninety for the males. These can best be described as brush-like tentacles that completely cover the mouth of the animal. The tips of the nautilus's tentacles are equipped with a taste organ. Nautiluses, which incidentally make up the oldest genus of the cephalopods, are the only living cephalopods that have a true outer shell, which serves both to protect them and to act as a flotation device. There is one other genus within the Octopoda order that has retained a shell. These are called paper argonauts (*Argonaut* sp.), named after the Greek word for sailor (*nautes*). Before they have reached their full size, female argonauts secrete a paper-thin detachable eggcase that is used for reproductive purposes and on occasion also acts as a shelter.

In contrast to other molluscs and except for the nautiluses and the argonauts, the cephalopods lost their hard casings, although vestiges of them remain as rudimentary, rigid inner support structures in squid and cuttlefish. Squid have a body part resembling a sword or a feather that is known as a gladius, or more commonly called a pen. It is somewhat flexible and made of chitin, a fibrous substance composed of polysaccharides. Cuttlefish benefit from the presence of a hard but brittle piece of porous calcium carbonate that resembles a flat bone. Octopuses have no support mechanism at all. In a sense, it is possible to say that the soft-bodied cephalopods have made a trade-off, sacrificing the protection afforded by a heavy outer shell for the ability to move around quickly. This choice has enabled them to evolve into some of the world's most dangerous, most wide-spread, and most successful predators.

The cuttlebone, a sort of inner shell in cuttlefish, is composed of calcium carbonate together with small amounts of mineral salts. It is sometimes crushed and used as a food supplement for poultry and caged birds. Cuttlebones are often found washed up on the beach and at one time country folk living on the coast of Jutland in Denmark thought they were scales from a whale.

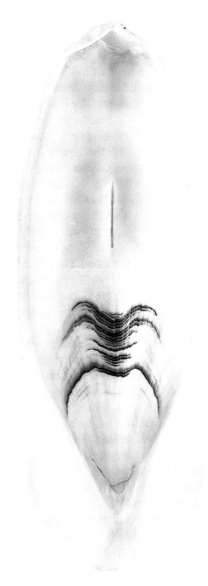

◘ Cuttlebone, a 24-centimetre-long inner support structure from a *Sepia officinalis*.

When Did the Cephalopods Appear on the Scene?

Mollusca, the second largest phylum of invertebrate animals, emerged under what is referred to as the Cambrian Explosion about 542 million years ago. During a period lasting for fifty million years there was an amazing outburst of new life forms, among them all the major phyla of animals of the present age. The cephalopods appeared toward the end of this time, probably still with outer shells. They later evolved to become some of the most successful and dominant marine creatures.

◘ Chitin pens from squid (*Loligo forbesii*).

Belemnites, named after the Greek word *belemnon* meaning dart or javelin, are a group of cephalopods that thrived during the Cretaceous Period (145–66 million years ago), but became extinct at the same time as the dinosaurs. While they resembled present-day cuttlefish, they had ten arms of equal length and no tentacles. At the rear of the animal there was a rostrum, a long, heavy calcite extension of their internal shell. These rostra are typically torpedo-shaped and eight to ten centimetres long. They are often found in fossilized form and are known popularly as 'thunderstones' because early Europeans thought they were formed when thunderbolts hit the ground.

The last of the common ancestors of both the nautiluses, with their hard shells, and the soft-bodied cephalopods lived about 416 million years ago, at the conclusion of the Silurian Period. Up to that point these animals had a comfortable existence, standing virtually on their own at the top of the food chain. But this all changed with the proliferation of fish life, which started with the Cambrian explosion and reached its heyday in the so-called 'age of fishes,' the Devonian Period (415–359 million years ago). From that time forward, cephalopods faced stiff competition for the same food supply. A few million years later, some of the ray-finned fish developed a strong jaw that was able to crush shells. The predators had themselves become very desirable prey.

It was probably because of this evolutionary pressure from fish that the soft-bodied cephalopods (Coleoidea) appeared in the course of the Permian Period about 270 million years ago. These had shed their outer hard shell allowing them to move around more easily to evade predators, but they had also become more vulnerable. When the teleosts, extremely successful ray-finned fish with a vastly improved jaw structure that could grab prey and draw it into their mouth, emerged during the Triassic Period (250–200 million years ago), the cephalopods' unchallenged mastery of the oceans came to an end.

◘ Thunderstone. Fossilized rostrum of a belemnite, a now extinct marine cephalopod. The fossil is formed when the hollow part of the dead animal's rostrum fills up with sand. Over time the rostrum disintegrates, leaving behind only the hardened silicon dioxide (quartz).

Some of the species of cephalopods that had shed their outer shell survived by becoming even more mobile, enabling them to cover great distances rapidly. They also adapted themselves to live at different sea depths everywhere in the world. Most probably to escape capture, those cephalopods that preserved their shell were forced to descend to deeper waters. Apart from this bodily transformation, the cephalopods evolved as the only group of molluscs with a brain. It is thought that the cephalopods, thanks to their diversity and to having both external and internal shells, but not especially due to their size, were among the most dominant forms of life in the oceans until sixty-five million years ago. The struggle with the finfish for supremacy in the seas persists to this day, but there are now indications that climate change may be upsetting this delicate balance.

The overall fossil record for Cephalopoda goes back as far as 296 million years ago and includes about 17,000 named species, compared to only about 800 identified species that have survived to the present. But there are relatively few examples of octopuses, squid, and cuttlefish. This is because their bodies are composed mostly of soft tissue and the hardest parts, the squid pens, are made of chitin, which does not fossilize nearly as well as calcium-containing bones or as the shells of extinct species. A fossil of an octopus that is about ninety-five million years old shows that these animals have undergone virtually no change since then.

Where and How Do They Live?

With the exception of the Black Sea, cephalopods inhabit marine environments all over the world. Generally they are found in greater abundance close to the continental landmasses. They live in both cold and warm waters, although there is a greater diversity of species in tropical zones. None of the cephalopods can survive in fresh water and they are critically dependent on the salinity level in the sea water. Some live under extreme conditions, for example, the incredible pressure several thousands of metres below the surface or the hot areas near underwater, volcanic seabed fissures. Some spend their time hunting for prey close to the surface, while others remain hidden at great depths at the very bottom of the sea. The various species of cephalopods are usually present all year round in their particular habitats, although some descend to slightly lower depths in winter and come nearer the coastlines in spring and summer when their eggs are hatching. They reproduce quickly because the fertilized eggs of many species rise to the surface and the hatchlings, which resemble the adults but are as transparent as glass, are widely dispersed by the sea currents, much like plankton. Some cephalopods lay as many as 100,000 eggs.

All cephalopods are predators, with the exception of nautiluses that are also scavengers. Octopuses eat anything that is alive and are not hesitant to wrap themselves around prey that is as big as they are. Their favorite foods, however, are shellfish, especially crabs.

Perhaps Ringo Starr was on to something when he wrote the lyrics to a popular Beatles song, *Octopus's Garden* …

"I'd like to be under the sea in an octopus' garden in the shade.

We would be warm below the storm in our little hideaway beneath the waves.

Resting our head on the seabed in an octopus' garden near a cave.

We would sing and dance around because we know we can't be found."

Cephalopods grow quickly and the growth rate is limited only by their relatively short life span, typically up to about two or three years. Some of the small species live for only a couple of months, although the nautiluses, which are the longest lived, can reach an age of fifteen years. Most types of cephalopods die after they have reproduced; the male immediately after mating and the female after laying the eggs and, in some cases, looking after them until they have hatched. The huge Pacific octopus *Enteroctopus dolfleini*, which has arms six metres long, lives for only two years and the giant squid *Architeuthis* rarely lasts for more than three years.

In the depths of the sea. The greatest known depth at which a living cephalopod has been found is currently 7279 metres. In this case, it was a *Grimpoteuthis* sp., known popularly as a Dumbo octopus because of its prominent ear-like fins that resemble those of the cartoon elephant. It is thought that generally only the very small octopuses live in such deep waters.

Squid often live in large groups that chase around in the open seas, snatching their prey with their long tentacles. The Humboldt squid (*Dosidicus gigas*) are some of the fiercest hunters. They sometimes attack in swarms that appear to behave in coordinated fashion, eating any living thing in sight. Octopuses, on the other hand, are mostly solitary and crawl around on the seabed where they catch their food using their strong arms and suckers.

There is much we do not know about cephalopods, especially about those that live in very deep waters. For example, we have no idea whether cephalopods sleep, what kind of communication there must be between them when they hunt in a pack, nor even why most species reproduce only once before they die.

Cephalopod Anatomy

In this chapter we will look at how the anatomy of cephalopods is organized, bearing in mind that in many instances the details vary from one species to another. Like humans, cephalopods have bilateral body symmetry, meaning that if they were split in half lengthwise, the result would be two pieces that are approximate mirror images of each other. While these creatures have many organs and structures that are analogous to those found in vertebrates, the exact way in which these function can be quite different. And, above all, the strange outward appearance of the cephalopods, the way in which they move about, and their ability to change their shape are a source of endless fascination.

Knowing something about how the anatomy of cephalopods has evolved, as well as their behaviour, is central to having an understanding of their place in the ecosystem, how they can be harvested sustainably, and why they are especially suitable for human consumption. This lays the groundwork for our later exploration of those edible parts of their bodies that are of gastronomic interest and how they can be prepared.

As the nautiluses are not an important food source, they are not particularly relevant to our discussion from this point onward and the term cephalopods is used in a general way to refer to octopuses, squid, and cuttlefish only, unless specified otherwise.

Mantle

The best place to begin is with the central muscular structure called the mantle. Because it makes up the central part of the animal it is sometimes misleadingly referred to as the body or head. The mantle is particularly well-developed in cephalopods and encloses the innards, digestive system, ink sac, and internal organs such as the three hearts, liver, reproductive organs, and gills, as well as the vestigial rudimentary inner shell found in some species. On octopuses the mantle has partially fused with a head-like area where the eyes are located and to which the arms are attached, while on squid and cuttlefish it is connected to a separate head.

The unfilled space inside the mantle, referred to as the cavity, acts as a flexible storage area for the water that has been drawn into it by expansion of the mantle muscles. This reservoir serves a dual function by ensuring that there is a fresh supply of water to pass over the gills or to be used by the siphon, which can also squirt it out with great force. The latter makes it possible for some species of cephalopod to move in a jet-propelled fashion, enabling them to accelerate more quickly than most fish and making them the speediest of all invertebrates.

When water is expelled from the cavity this action sets up a tension in the mantle that can be relieved by drawing in new water. Squid can empty the mantle cavity so efficiently that almost 95 percent of the liquid is expelled in a single action. This indicates that the mantle is sufficiently elastic to allow the volume of water in the cavity to increase or decrease by a factor of twenty. The mantle muscles in squid and cuttlefish are organized differently from those in octopuses, which has a great impact on their ability to swim. The outermost part of the mantle of squid and cuttlefish has a particularly dense structure so that less energy is expended on swimming.

Nerves, Brain, and Intelligence

Nerves, which developed very early on in animal evolution, are structured in the same way in all living organisms that have a nervous system. In those organisms that have a brain, however, its structure varies from one animal to another. What distinguishes the mammalian brain is the presence of a neocortex that deals with the brain's more complex functions, such as cognition, sensory perception, planning, speech, and rational behaviour. Having a neocortex is, therefore, often associated with intelligence, although it may not be the only determining factor. Cephalopods have no neocortex, but their brains have folds and a complex morphology.

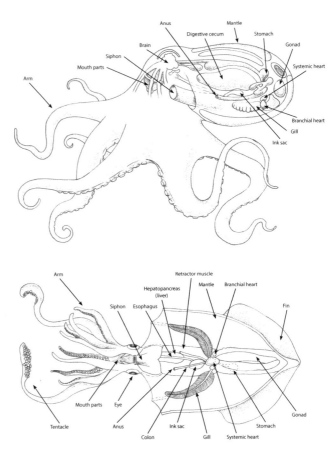

◨ Schematic illustrations of the internal structure of an octopus and a squid.

The human brain and nervous system together make up a hierarchical structure that, apart from certain reflex actions, functions as a top-down system. The cephalopod brain is quite different and has both central and distributed components. The central portion, which is protected by a cartilaginous skull, wraps around the esophagus, which passes right through its middle. On each side, the brain bulges out into a large optic lobe that is connected to an eye, which is an integral part of it. What is very unusual about the cephalopod nervous system is that it is much more decentralised than the human one and about three-fifths of the brain is located in the arms and tentacles. As a result, the arms are somewhat autonomous, can make decisions locally, and can carry out functions independently of each other. Nevertheless, they can still exchange information without going through the central brain, allowing them, as if by a miracle, to carry out very coordinated operations. It is not clear precisely how the arms work with each other.

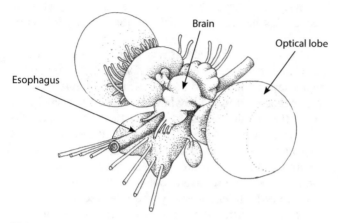

Brain

Optical lobe

Esophagus

□ Schematic illustration of an octopus brain, optic lobes, and eyes. The esophagus passes through the middle of the brain.

The nervous system found in each cephalopod arm exhibits a behaviour called autonomy. An octopus is able to break off and jettison an arm, which can then move about independently for a short time without being controlled by the central brain. This is useful in a fight against a predator as the arm still looks alive and can act as a decoy to draw the attacker's attention away from the rest of the octopus, allowing it to escape.

An important characteristic of an animal's brain is the relationship between the weight of the brain and that of the body, known as the brain-to-body ratio. Along with the degree of connectivity between the neurons in the brain, this relationship determines the brain capacity and possible intelligence level of an organism. For example, this is what separates humans and dolphins from other animals of the same size. We have a brain that makes up about two percent of our body weight, while other animals of the same size have brains that are ten times smaller.

Octopus vulgaris can regenerate an arm that has been lost with a completely new one in about four weeks.

San-nakji—**live octopus**. This dish is a Korean specialty made from pieces of an arm that have been cut from a small, live octopus. It is usually seasoned with sesame seeds and sesame oil and served while the pieces are still able to move around. They continue to move because the autonomous nervous system is still intact, even though the animal has died, the hearts and the gills have stopped working, and the central brain has shut down. Given that the suckers can also still be active on the severed arms, one should eat this dish very carefully as they can stick to the inside of the mouth and throat and cause choking.

The body-to-brain ratio of octopuses falls somewhere between that of fish and birds and it is possible to imagine that in some ways they might be as intelligent as birds. Nevertheless, there is an indication that the common octopuses, *Octopus vulgaris*, are much smarter than most birds. Some cuttlefish from the *Sepia* genus also have a relatively large brain, but we know even less about what impact this has on their existence than we do for octopuses.

One can speculate about why cephalopods, and octopuses in particular, have such relatively large brains. This might have something to do with what distinguishes them from other molluscs. They have no hard outer shell into which they could retreat to protect their soft bodies. So in order to survive in their challenging three-dimensional environment they had to develop the ability to move rapidly and navigate skillfully through their surroundings. In addition, they had to master complex, coordinated movements in order to capture prey or evade predators.

Squid and cuttlefish have somewhat smaller brains than the octopuses. But while one might say that the octopuses are the smartest, the other two are the fastest. This has to do not only with their muscle structure but also with their nerve fibres (axons), which are able to transmit impulses incredibly quickly. They are so thick that they have been used to formulate the scientific basis for our overall understanding of nervous systems and brains, including those in humans.

The Giant Neuron

It is generally accepted that although organisms may differ greatly from each other, their neurons basically function in the same way. Their job is to receive signals from other neurons through branched extensions called dendrites located on the cell body. They then process the signals and send them on through thin fibres called axons that, in turn, pass them on to adjacent neurons via the synapses. Each neuron is a single component of the whole neural network that makes up the brain.

The difficulty in studying individual neurons and their function is that the axons in the majority of organisms are very thin. In the case of humans they are fifty times thinner than a hair, whereas cephalopod neurons are much thicker. The axons in special nerves of the squid *Loligo pealei* measure 0.5–1.0 millimetre in diameter and are several hundred times thicker than human ones. It is, therefore, much easier to remove them intact from that organism in order to investigate their electrical properties.

As a consequence, the axons from *Loligo pealei* have become the most frequently used specimens for carrying out research on nervous systems. Much of what we know about the workings of human nerves is based on such studies. The groundbreaking experiments were carried out by Alan Lloyd Hodgkin and Andrew Fielding Huxley, who were able to identify the mechanisms that govern how nerve signals are transmitted via electrical impulses through special ion channels in the membranes of the neurons. For this seminal discovery they were awarded the Nobel Prize in Medicine in 1963. In recent years researchers have again turned to the giant *Loligo paelei* axon to gain further insights into different molecular transport mechanisms within the neurons, work that may help to explain the causes of various nerve-related diseases. It is also possible that by studying the ability of cephalopods to regenerate nerve cells, for example, after they have lost and regrown an arm, one might learn something about how to repair damaged human nerve tissue.

Do *Loligo pealei* and other species with similarly large axons derive any benefit from this characteristic? Absolutely–the thicker the axon, the more rapidly it can transmit electrical signals. In the case of *Loligo pealei*, for instance, nerve signals can move through their body and arms at the amazing speed of 100 metres per second, which is ten times faster than they can move through human axons. As a result, they can swim extremely rapidly, as if jet-propelled.

Is an Octopus Intelligent and Does It Have Consciousness?

Ninety-five percent of all animal species are invertebrates. When one compares an octopus with an oyster it seems glaringly obvious that it must be more intelligent, especially when one looks at the evidence. It can build a den complete with a garden-like exterior, use tools, unscrew the tops of bottles, untie knots, open safety clasps, make plans, recognize certain people, and find its way in a labyrinth. It can exhibit a great deal of adaptability in the way it behaves and it appears to retain a significant amount of knowledge and to have a good memory. It is curious and explores its surroundings using its mouth, arms, and suckers. It is able to figure out how to solve problems on its own but also by observing how other octopuses tackle them. In addition, it reacts to identical external stimuli in different ways depending on the context. It is, therefore, not without reason that the octopus is said to be the most intelligent of all invertebrates.

One might be tempted to wonder whether it was ever intended that cephalopods should have such a large brain. At the very least, it seems inconvenient that the esophagus passes right through the middle of the brain. There are also examples of cephalopods choking on a fish bone stuck in their gullet, but in this case it is actually lodged in the middle of their brain.

A large octopus has approximately 350 million brain neurons, a number that is only about 300 times smaller than in a human, about 20 times more than a frog, and 30,000 times more than a snail, which also belongs to the Mollusca order. The brain of a cat is about twice as large as that of an octopus. Given this background, it would appear that there is an underlying basis for the behaviour of an octopus that is not linked solely to the comparative size of its brain. The number of neurons *per se* is not the whole story, as intellectual capacity is also governed by how they are connected. In addition, it is doubtful whether we have any way to gain insight into how cognition functions in a being that is so very different from us. And there is always the danger inherent in anthropomorphizing aspects of the world around us, for example, when we interpret the activities of an octopus as 'play,' its particular behaviour as evidence of its personality, or interpret its solitary existence as a sign that it is antisocial.

◘ Under the watchful eye of an octopus.

Many who have observed an octopus through the glass wall of an aquarium, not least among them researchers working with these creatures, have had their doubts as to exactly who was looking at whom. Is there some reason for, or a thought process controlling, that sharp, focused gaze of the octopus?

Our inability, or possibly reluctance, unequivocally to attribute true intelligence to the octopus is perhaps just as much an indication that our way of thinking may not stretch far enough to acknowledge and understand forms of intelligence other than our own. When viewed from that perspective, Aristotle's assertion that the octopus is "a stupid creature" seems to make more sense. Nevertheless, studies of octopuses could help to expand our own understanding of the nature of intelligence.

Even though there is considerably less knowledge about brain function in squid and cuttlefish, research has shown that, like octopuses, some of them have spatial learning capacity. For example, *Sepia* demonstrate this by being able to move through a labyrinth. *Sepia* also exhibit individual behaviour patterns and appear to be much more sociable than octopuses.

"Anyone who has seen an octopus in an aquarium will have had the uncanny impression of being carefully watched."

Roger T. Hanlon, octopus researcher (1996)

"By studying the octopuses' intelligence, play, and personality, we can look at aspects of ourselves from another angle."

Jennifer Maher, octopus researcher (2003)

A Consciousness of Another Kind

In his book *Other Minds: The Octopus, the Sea, and the Deep Origins of Consciousness* (2016), the Australian philosopher of science Peter Godfrey-Smith takes up the challenge of exploring the relationship between sentience, intelligence, consciousness, and the physical world. His inspiration comes from the observation that there were two seemingly independent evolutionary paths that led to two very distinct species, humans and octopuses, which each have their own advanced, but very different brains. He thinks that the octopus is "probably the closest we will come to meeting an intelligent alien." Furthermore, he suggests that we should think of this as an opportunity to gain a better understanding of what is meant by intelligence, which in turn could influence how we might conduct ourselves on encountering another being that, like us, has a mind and is aware of its own existence.

A fundamental task in the field of philosophy is to research the connection between the mind and matter and, more specifically, the question of how a bodily construct such as the nervous system and the brain can shape what we call a mind. In the past half-century, this age old problem has taken on increasing importance with the advent of computers. In many ways these are a sort of 'electronic brain' designed to mimic a neural network resembling our own that is able to learn, has memory functions, and possibly also possesses intuition.

Computer and communications systems, including the internet, are undergoing constant change, working at ever increasing speeds, becoming more complex, much, much larger, and more integrated into local and global infrastructures. Apart from that, they have a degree of autonomy, being able to regulate, correct, and reprogram themselves. This naturally gives rise to speculation about whether such systems could be the genesis of a form of intelligence with a mind of its own, exactly like the computer Hal in Stanley Kubrick's 1968 movie, *2001: A Space Odyssey*.

In order to appreciate the possibility that there are different types of intelligence and, consequently, minds, Peter Godfrey-Smith takes us on a journey back to what is called the last common ancestor of humans and cephalopods, to the time when their evolutionary paths diverged about 600 million years ago. This was long after the appearance of the first forms of life, which is thought to have taken place at least 3.8 billion years ago, a mere 700 million years after the Earth came into being. Some of these eventually took on the shape of unicellular organisms that could sense and react to external stimuli, partly from the presence of food sources and toxins, but later also from signals coming from others like themselves. They reproduced by cell division. Most likely the transition to multicellular animals came about because a cell did not divide fully and the two parts remained attached to each other. When this process repeated itself several times, it formed a ball-like mass of cells organized and controlled by their interconnected signals. In the course of evolution, the cells differentiated to carry out specialized functions, advances that collectively resulted in a better chance of survival for this proto-multicellular organism.

The last common ancestor originated from a population of these slightly more advanced animals. To get a mental picture of it we can visualize it as a little flat worm-like creature that, like all forms of early life, lived in the sea. This organism was bilaterally symmetrical and it might have had simple eyes or patches sensitive to light. It also had the benefit of a simple nervous system, which helped it to sense its surroundings and to send signals from one end of its body to the other. Nervous systems had actually developed in an earlier evolutionary period and were also already present in some other soft-bodied animals such as jellyfish, comb jellies, and sea anemones, but were absent in sea sponges.

We now arrive at the time 600 million years ago when there was a fork in the evolutionary path of the common ancestor. One path led through many branches to two phyla of invertebrates—the arthropods, for example, insects and shellfish, and the molluscs, which include the cephalopods. The other path led to the vertebrates, such as fish, birds, mammals, and ultimately to humans.

From this point of divergence, the cephalopods evolved in dramatically different fashion from both the other invertebrates and the vertebrates. Alone among the invertebrates, they developed complex nervous systems, but in a way that was completely unlike those found in mammals, including humans. Whereas the human brain has become a large concentration of nerve cells in one place in the body, the cephalopod brain is more decentralised, with most of it distributed in the arms. To the extent that an octopus can be thought also to have a mind, it appears that the evolutionary process resulted in two very different models of sentient beings. That naturally leads to the question of how unique humans are as a species. When we search for intelligent life elsewhere in the universe, we should perhaps prepare ourselves for the probability not only that we will find it, but also that we should not be imagining it in the shape of stereotypical small green humanoids with whom we can communicate, even if with difficulty.

The evolution of advanced nervous systems and a brain-like structure made up of nerve cells can be interpreted as a sign that the various organisms present at the end of the Cambrian Explosion about 600 million years ago began to relate to each other in new ways. The acquisition of distinguishing characteristics such as hard shells, claws, eyes, fins, legs, and antennae indicate that life had emerged from a sheltered environment, a sort of paradise, into a world where tools were required both for protection and to hunt for food. No longer free to ignore each other, the animals now evolved as a consequence of their behaviour and the mechanisms for controlling it. In the words of Peter Godfrey-Smith, "From that point on, the mind evolved in response to other minds."

Eyes

Squid, cuttlefish, and octopuses all have prominent eyes that are very large in proportion to the rest of their bodies. The biggest are those of the colossal Antarctic squid, which can be up to twenty-seven centimetres in diameter and are the largest in the world.

Although the actual size of the eyes is a very striking aspect of the sense of vision in cephalopods, it is worth noting that it is in some ways more sophisticated than in humans. Like us, they have what is called camera-type vision and images are formed on a retina via a lens. But their optic nerve fibres are routed behind the retina, rather than in front, and they, therefore, have no blind spot. They are, however, colour blind because they have only one kind of light receptor in the retina, whereas humans have three and can distinguish between the three basic colours, red, blue, and yellow. As a result, cephalopods see their surroundings in shades of grey, with colours in the green end of the spectrum being most intense. Unlike humans who can barely see polarized light, cephalopods can detect differences in polarized light and orient themselves in relation to it. Their vision is usually monocular, which is to say that they see with one eye at a time. The resulting visual impressions are not coordinated, as they are in humans, and each eye, therefore, forms its own picture.

The pupils in some of the cephalopods are different from the circular ones in humans. Those in cuttlefish are in the shape of a W, while those in octopuses have a rectangular, narrow slit that is automatically adjusted to remain in the horizontal position, resulting in a very sharp image in that plane.

▢ Eyes and pupils of *Loligo*, *Octopus*, and *Sepia*.

Mouth and Beak

The oral cavity of cephalopods is a muscular mass that contains the mouth, beak, and raspy tongue (radula) and serves as the entry point to the digestive system. The beak is made up of chitin, which is hard, and resembles that of a parrot, although with an underbite. It is as sharp as a razor blade and can slice its prey into small pieces in a matter of seconds. The ribbon-like radula has rows of tiny, spiky teeth that are also extremely sharp. It can drill a hole in the shells of prey and reduce the food to bits. This structure is most highly developed in squid and cuttlefish.

◨ Beak and radula from a cuttlefish (*Sepia officinalis*).

The posterior salivary gland, located under the radula, produces viscous secretions that are transported through ducts and discharged into the oral cavity. These contain venoms and certain enzymes and can be used to paralyze and break down prey. For example, the enzymes can separate the flesh from the shell of a crab or a mussel so that the cephalopod can suck it out. The toxin in the venom in some cephalopods is produced by microorganisms that live in the venom gland. But bites from cephalopods are not particularly harmful to humans, although they can provoke an allergic reaction. Only those from small blue-ringed octopuses, *Hapalochlaena* spp., are dangerous and can even be fatal.

Arms and Tentacles

When we think of cephalopods what almost always springs immediately to mind is an image of those amazing, sometimes frightening looking appendages—their arms and tentacles. They are multipurpose and are used for hunting, locomotion, and reproduction. Octopuses have four sets of arms, while squid and cuttlefish have four pairs of arms and one pair of tentacles, always in a bilateral symmetrical arrangement on the body. What is unusual about these appendages is that they are attached radially around the mouth rather than directly to the main part of the mantle. In a sense, they function as a sort of extension of the oral cavity or pharynx and enable the cephalopod to catch and hang on to larger prey until the beak has a chance to cut it to pieces that can then be moved on into the digestive system.

Arms and tentacles make up a significant proportion of the whole animal. Half of the entire volume of an octopus consists of the muscle mass in the arms. They can easily be stretched to twice their length, but this is nothing compared to the tentacles of squid and cuttlefish as they can be extended to several times their length when at rest and often make up the greater part of the extent of the animal. This elasticity has made it very difficult to determine and describe the size of squid and cuttlefish in a definitive way. For example, the few giant squid tentacles that have washed ashore could be stretched to a length of up to ten metres.

Radula—the key to an important recent discovery. When working at Yale University on some rock samples that had been collected in Morocco about ten years previously, an international group of scientists found what they had overlooked before. They were able to identify the imprint of a radula in the fossilized remains of an early organism. The fossil was less than two centimetres long and showed the traces of an animal that lived 480 million years ago. The researchers concluded that it definitively had the characteristics of a simple mollusc and that it was possibly the very early ancestor of all the molluscs that came after it.

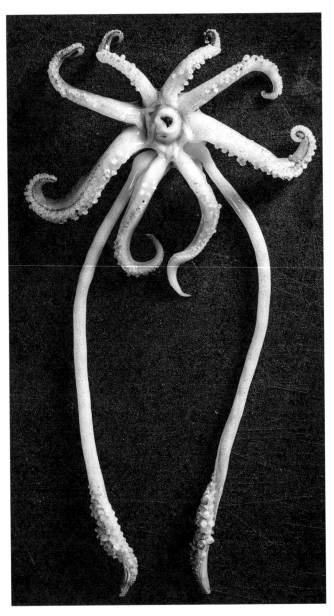

■ Arms, tentacles, beak, and buccal mass from a squid (*Loligo forbesii*).

The arms of an octopus move relatively slowly, at a speed of about two kilometres per hour. But they are especially strong, being able to pull with a force that is equivalent to hundreds of times the body weight of the whole animal, and have suckers that can hold the prey.

The tentacles of squid and cuttlefish are longer than their arms and are deployed on their own as hunting tools. They are often shaped like a club with a thin stem that has only a few small suckers, ending in a larger head with more well-developed suckers and small sharp chitin hooks, arranged singly or in pairs. Sometimes the tentacles are tucked away into pouches near the mouth, but can suddenly be shot out because their muscles act incredibly rapidly. The tentacles are made up of fast muscles that can propel them toward prey at great speed. For example, a cuttlefish is able to grab onto its victim in a matter of about 1/100th of a second. The velocity at which these appendages can move through the water can be up to 900 kilometres per hour, which is ten times as fast as one can throw a ball or a spear and approximately the same speed as that of a jet plane. The giant squid (*Architeuthis*) does not move as quickly as the smaller squid species, but has certain advantages in terms of capturing prey. The suckers on its arms and tentacles are lined with sharp serrated chitin edges that can drill and cut into prey, and there are several clubs with large suckers at the end of tentacles.

Arms and tentacles are also important for locomotion. Those on squid and cuttlefish may act as rudders when they are swimming. Octopuses are even able to 'crawl' on them on the seabed and can bend them at will in all directions. This allows them to turn the sides with the suckers toward prey to draw it close to them or to feel their way around their immediate environment.

Cephalopods have a modified arm called the hectocotylus, with a particular organ (ligula) at the end that can transfer a protein capsule containing sperm (spermatozoa) to the female during mating. In the case of octopuses, these are inserted into the mantle of the females, after which the arm is sometimes wrenched off and left in the female. Unlike arms that are lost in a struggle with a predator, the hectocotylus cannot be regenerated. Male squid and cuttlefish can deposit sperm capsules just about anywhere on the body of the females, but more commonly under the mantle or near the mouth parts. There are examples of the spermatozoa rather brutally boring into the flesh of the females and lodging there, being set free only when the eggs are to be fertilized.

Suckers

Even though it is a huge animal, the giant squid *Architeutis dux* is not especially strong or fast moving in comparison to other cephalopods with ten appendages. This is partly because the flesh in its mantle contains a great deal of ammonium chloride that, being lighter than seawater, helps it to regulate its buoyancy. In addition, its nerve fibres are much thinner, approximately 250 micrometres versus 800–1000 micrometres for some species of *Loligo* squid, and consequently they are unable to process signals as quickly. On the other hand, *Architeutis* has a large beak and extremely long tentacles.

Most cephalopods have suckers on their arms, with those on octopuses being especially well developed. The suckers are controlled by their individual nervous system and, as a consequence, can be activated independently of each other. Each sucker has about half a million nerve cells that in essence make up a mini-brain.

The suckers are a formidable tool for catching prey, grabbing on to it, and holding it tight until it has been paralyzed, bitten, killed, and cut up into small pieces by the beak. It is then ground into even smaller pieces by the raspy tongue, and disappears through the mouth into the digestive system. Suckers have evolved differently depending on the species. Those on octopuses are attached directly to the surface of the arms, whereas those on squid and cuttlefish are placed on top of a small stalk. In addition, the suckers on the latter two can have about twenty-five tiny, sharp chitin teeth just inside their edge. These help by sinking into the surface of the prey and hold it even more firmly.

How do the suckers work? The muscles that are attached to the edges of the suckers can create a considerable negative pressure when there is a tight seal between them and the surface of the animal, much like a suction cup but considerably stronger. A large sucker has a load-carrying capacity of up to fifteen kilograms, which is dependent on the internal volume of the sucker and on whether water or air is trapped under it. Because the cohesion of water is greater than that of air, the sucker can exert more force on a wet surface.

◘ The club-like end of a tentacle on the squid *Loligo forbesii*. The edges of the suckers are ringed with tiny chitin 'sawteeth.'

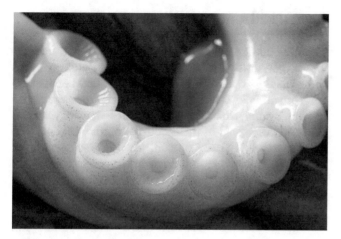

■ Suckers on the octopus *Eledona cirrhosa.*

The suckers on squid and cuttlefish are able to generate a greater force than those on octopuses. But octopus suckers have a more advanced structure, are more flexible, and due to their well-developed nervous system and the muscles at their edges, they are able to adapt to function on very uneven surfaces. On an octopus the individual suckers can move around and rotate individually. They can also be pushed away a little from the arm and even fold up so that they act as a sort of tweezer. This enables octopuses to use their arms to perform very complex tasks such as tying knots or moving an object along their length by transferring it from one sucker to the next. The largest suckers on a big octopus are about eight centimetres in diameter, somewhat larger than the five centimetre ones on the giant squid *Architeutis.*

Apart from their ability to carry out mechanical tasks, the suckers are also equipped with a number of sensory receptors. An octopus can actually taste and feel with its suckers and in that way check out its surroundings and find food sources. This form of sensory perception is carried out by a combination of special receptors on the edge of the suckers. They are able to detect chemical signals, measure mechanical pressure and the effects of force, and convey information about the position and movement of the body. As an *Octopus vulgaris*, for example, has about 240 suckers on each arm and each sucker has approximately 10,000 receptor cells, it is obvious that its sensory abilities are exceedingly advanced and much more finely tuned than our own. Unlike the octopuses, the cephalopods of the *Sepia* genus have not developed the ability to taste with their tentacles and instead depend on their eyes to locate food.

Can octopuses hear? It is not known whether octopuses have a true sense of hearing, but they are able to detect vibrations in the range of 400–1000 Hz.

Blue Blood and Three Hearts

Cephalopods have blue blood, both when it is transporting oxygen and when it is deoxygenated, although it is somewhat less opaque and lighter in colour when its oxygen content is low. The blue colour is due to the presence of hemocyanin, which relies on copper to bind oxygen and is directly dissolved in the bloodstream and not encapsulated in a cell. By way of contrast, the iron content of hemoglobin, which is found inside the red blood cells of human blood, is responsible for the red colour of oxygenated blood. Other invertebrates, for example lobsters, also have blue blood.

The differences between how hemocyanin and hemoglobin function have important implications for cephalopods, conferring both advantages and drawbacks. As copper has significantly lower oxygen binding power than iron, cephalopod blood has three times less capacity to transport oxygen. Consequently, they have less stamina than organisms such as fish and mammals that depend on hemoglobin. On the one hand, hemocyanin can still function where the oxygen concentration is really low, giving it an advantage over hemoglobin, which is much less effective in such environments. We are familiar with this problem from the shortness of breath we experience when we try to exert ourselves at high altitudes. On the other hand, the oxygen carrying capacity of copper is sensitive to the acidity level in the blood, which renders the cephalopod respiratory system less robust. This means, among other things, that the acidification of the oceans due to increasing quantities of dissolved carbon dioxide caused by climate change may have important repercussions for the proliferation and distribution of cephalopod populations.

Cephalopods have three hearts, arranged in a row. The one in the middle, called the systemic heart, pumps oxygenated blood to the organs. The two outer ones, called branchial hearts, move oxygen-depleted blood through the delicate feather-like gills. The circulatory system in cephalopods is well developed and they are the only molluscs that have a closed system. It is, however, not especially efficient. For example, the systemic heart of octopuses shuts down when they are in motion, which means that the animal tires very quickly. Its energy requirements are particularly high when it is squirting water out through the siphon to swim by jet propulsion.

Muscles

In order to be able to understand the particular nature of cephalopod musculature it would be useful to take a look at what is required to enable an animal to move about. Every locomotor apparatus needs some sort of support structure against which to move. Animal groups have solved this problem in a range of distinctly different ways.

Vertebrates such as fish, birds, and mammals are supported by an inner skeleton, while arthropods such as crustaceans and ants have an outer one. The limbs are constructed like a set of linked levers that can be moved in relation to each other around axles (joints). Power to effect movement is delivered by the muscles that are fastened to the levers. The disadvantage with such a system is that movement can take place only in the directions that are permitted by the joints and in distances that are determined by the length of the levers.

A Muscular Hydrostat

Invertebrates, including the cephalopods, evolved in a way that solved the problem posed by the lack of a support structure. The solution, which is known as a muscular hydrostat, not only compensates for their lack of a skeleton, but also left them with an enormous amount of freedom with respect to how they move, in which directions, and over what distance.

A hydrostat is a deformable system that is subject to constant pressure. If, in addition, the system is made up of material that cannot be compressed, it follows that the system also has a constant volume but may change shape due to forces acting on it. Muscles are hydrostats as they consist mostly of water, which is not compressible under conditions of normal pressure. The human tongue and the trunk of an elephant are good examples of this type of muscular hydrostat.

One can think of a muscular hydrostat as being like a balloon that is filled with water, but not fully distended. If one presses on the balloon in one place, it has to bulge out in another in order to maintain constant pressure and volume. Similarly, if one pinches it to form a pipe shape in one place, it will contract in the other direction. It is this simple condition of constant pressure accompanied by the possibility of changing its shape that enables cephalopods to move in the way that they do.

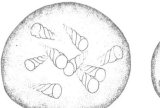

◘ Schematic illustration of the working of a muscular hydrostat. The shape of the balloon on the left changes and bulges out to look like the one on the right, but its volume remains constant. The twisted spirals inside the muscle represent collagen fibres.

Squid and cuttlefish have a rudimentary inner support that is a vestige of the external shell of the animals from which they evolved, but it is of little importance for locomotion. As mentioned earlier, the inner support in squid takes the form of a flexible, feather-shaped structure called a pen. Cuttlefish have a broad, porous inner shell, the cuttlebone, which acts as a flotation device. The octopuses' complete lack of support structures, as well as the flexibility of their muscles and the fact that they function like a hydrostat, means that the only thing that limits their ability to pass through holes and hide in crevices is the size of their beak and of their cartilaginous skull. Apart from these constraints, octopuses can move more or less like a viscous fluid and slip in and out of any opening. This 'talent' is not lost on aquarium keepers who know only too well that it takes a concerted effort, and possibly luck, to keep an octopus captive in a tank.

Cephalopod Muscles

Muscles are constructed hierarchically from bundles of muscle fibres, which are encased in connective tissue. The connective tissue itself is also a hierarchical structure, composed of collagen fibres, which are long protein molecules that are twisted around each other like a rope. These individual protein molecules are, to a greater or lesser extent, chemically cross-linked. The strength, toughness, and elasticity of the connective tissue increases with the number of cross-links and that is what determines how tender the meat can become when it is prepared for eating.

Collagen in finfish is much weaker than in cephalopods, which typically have four times more connective tissue than fish and also significantly less fat. In addition, the individual muscle fibres in cephalopods are considerably longer and generally ten times thinner than those of finfish and terrestrial animals. Among other things, this means that cephalopod muscles are tighter and smoother with a much finer structure. The exceptional strength of their connective tissue is what makes it possible for cephalopods to move rapidly and effectively with the help of the principle underlying a muscular hydrostat. If the muscle mass were more like an unstructured liquid without a significant amount of connective tissue, this would not work.

The three-dimensional organization of the muscle fibres in a muscular hydrostat allows the fibres to move in three different directions: parallel in a certain direction, at right angles to it, and twisted, sometimes like a spiral. The muscle fibres that run parallel to the length of an arm can, for example, shorten it or, if activated differently on its two sides, cause it to bend. Those muscle fibres that are perpendicular to the arm and tentacles can go crosswise in several directions. Another possibility is that the muscles spread out radially in all directions or are arranged in circles as in the mantle, as well as in the tentacles on squid and cuttlefish.

The movement of the muscles can cause a muscular hydrostat to change shape, even though the only way they can exert force is by contracting. Then other muscles, also contracting, can cause the first muscles to stretch out again. The locomotor apparatus of animals such as mammals is made up of striated muscles that can contract and relax in one direction only and, consequently, it is organized one dimensionally. In contrast to this, the muscles of cephalopods are organized three dimensionally and, as a result, they can support whichever way the arms choose to move.

Apart from the major muscle groups, there are many small muscles that govern such things as movement of the eyes and suckers and control the surface structure of the skin, as well as its colour and the patterns they form.

The incredible flexibility of octopus arms is due to their makeup. They are composed of a tightly packed network of muscle fibres and connective tissue that encases the central bundle of nerves. There are longitudinal ones that are parallel to the plane along the length of their arms, transversal fibres that lie in planes perpendicular to the length of the arms, and some that lie crosswise and wrap around the arms. The transversal fibres stretch out the arms, the longitudinal ones shorten them, and the crosswise ones cause them to twist. The arm is bent when the transverse fibres cause it to become thinner, while the longitudinal ones pull unequally on two sides of the arms, but at the same time keep its length constant. When a number of different muscle types interact with each other, they can change the stiffness of an arm.

How Cephalopods Move around in their Surroundings

Cephalopods have a number of ways to move around in their aqueous environment, and do so by swimming, using jet propulsion, gliding above the surface of the water, or crawling along the seabed.

Swimming takes different forms, depending on the species and how fast the animal wants to go. To swim slowly some octopuses use their arms in conjunction with the umbrella-like web that extends from the mantle to about a quarter of the way down the arms and acts like a swimming bell. Cuttlefish and squid swim by gently undulating the fins along their sides. But to move more rapidly, for example, when they are fleeing a predator, they all normally have their arms together like a tail. When they are chasing prey, squid and cuttlefish bring them together to a point in front of them in the direction in which they are moving.

In order to move really quickly, a cephalopod depends on jet propulsion. In this case, the mantle and the ability to alter its shape are the essential elements that enable the animal to jet propel itself by taking water into its cavity and then squirting it out again through the siphon in a certain direction. The muscle structure of the mantle, which in octopuses is different from that in squid and cuttlefish, is quite complex, with radial and circular rings of muscles that are vital for jet propulsion.

◘ Schematic illustration showing the different muscle structure in the mantle of an octopus (on the left) and of a squid (on the right).

Octopuses, which move quickly by this means, have thick mantles and strong circular muscles. When the circular muscles of an octopus contract, a set of muscles that lie parallel to the mantle also contract themselves. As a result, instead of the mantle elongating, its wall becomes thicker and water is expelled. This causes the radial muscles to be stretched out and water is drawn in again, which in turn relaxes the circular muscles. The circular muscles are now ready for a repeat operation and a new jet stream can be squirted out.

Squid and cuttlefish lack the parallel muscles that run the length of the mantle and, consequently, have a different mechanism for setting the jet stream in motion. Their mantle is covered with a special outer layer of muscles made up of a network of very strong collagen fibres, making it very elastic, much like a garden hose or a radial tire for a car. The mantle does not become longer when the radial muscles contract, which causes the water pressure inside it to rise. Because longitudinal muscles are absent in their mantle, squid and cuttlefish require significantly less energy for jet propulsion, turning some species into exceedingly fast predators. Some squid can sprint for short periods as fast as sharks at speeds of thirty to forty kilometres per hour.

When water inside the mantle is squirted out through the siphon with great force, it even enables some types of squid to break through the surface of the water and glide through the air, much like a flying fish. For example, the large flying Humboldt squid (*Dosidicus gigas*) can attain such a high speed that it has been known to fly through the air several metres above the water for more than ten metres.

Octopuses often crawl slowly on their arms by contracting an arm, placing it firmly on the ocean floor, and then stretching it out again and using suckers to push and pull themselves along. There are two species of tropical octopus that have even been observed to use bipedal motion to evade predators. They lift up six of their arms and walk backward on the other two, in that way disguising themselves as another marine organism such as a piece of seaweed.

The 'rhythm-less step' of the octopus. When an octopus crawls there is neither pattern nor rhythm to how they move their arms. They do not do it either left-right or right-left or one arm after the other in sequence—it is completely arbitrary and seemingly dependent on a last minute decision about which arm to use to go in a particular direction. This is different from the known locomotion patterns of every other animal, yet another way in which octopuses are unique.

Siphon

Cephalopods have a siphon, which is a muscular tube that allows water to pass from the mantle cavity into the surrounding water. The siphon is connected to a set of strong retractor muscles that encircle the innards. When the muscles contract. they are able to shoot water out through the siphon. The water acts like a jet stream that creates a force causing the animal to recoil in the opposite direction. Cephalopods are able to regulate the speed of the jet stream so that it results in both slow and fast movement, up to forty kilometres per hour. As the outlet of the siphon can be turned in all directions, this mechanism gives the cephalopod a great deal of freedom of movement, especially in comparison with an animal like a fish or an arthropod. Some squid and cuttlefish have a flap that can seal off the outlet of the siphon.

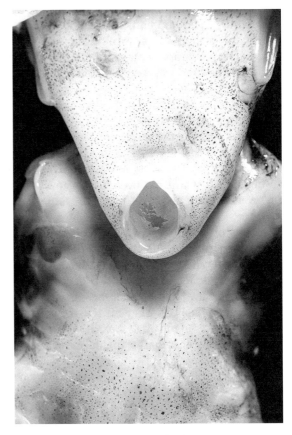

◘ The siphon from a squid (*Loligo forbesii*).

Apart from movement, the siphon is a handy multi-purpose tool. It is used to squirt out the cloud of ink for which cephalopods are famous and to blast away bits of the seabed and thereby possibly reveal potential prey or to dig up shells to build their dens. Last but not least, it serves a housekeeping function as it can be used like a water hose to tidy up the octopus's den or to clean its eggs.

Ink

In some languages, the popular names for cephalopods, for example, 'inkfish,' reflect their ability to squirt out a stream of dark liquid, known as sepia ink, which was the first type of ink used for writing in ancient times. It has the advantage that it can be diluted with water, producing a gradation of colours, ranging from very dark black to brownish tones. In fact, the word sepia now refers to a particular warm reddish-brown colour.

Apart from a few species of octopuses that live in such deep, dark waters that it would serve no purpose, all cephalopods produce a dark liquid, which is a mixture of ink produced in the ink sac and mucus secreted in the funnel organ. In octopuses the ink sac is located in the liver, whereas in squid and cuttlefish it is placed outside of, but near, the liver. The ink passes from the ink sac into the small funnel organ and the ink-mucus mixture then enters the siphon and is squirted out into the ocean.

The pen and ink connection. The familiar Italian word for squid, calamari, can be traced back to the medieval Latin word *calamarium*, meaning an 'ink pot' or 'pen case,' which itself ultimately stems from the Greek word for pen, *kalamos*.

The colour of cephalopod ink is due to eumelanin, one variety of melanin, which is the naturally occurring pigment that is also found in human hair and skin and determines their colour. Octopus ink is black, while that of squid such as *Loligo* is blue-black, and that of *Sepia* is brownish.

Cephalopod ink is derived partly from an amino acid, tyrosine, which has been broken down with the help of the enzyme tyrosinase. The presence of this enzyme can cause eye irritation and confuse the senses of taste and smell of an attacking predator. In this way, the ink can be considered to act as a toxin and, if sufficiently concentrated, it can actually adversely affect the cephalopod itself by clogging its delicate featherlike gills. In addition, the ink contains an abundance of free amino acids, as well

■ An octopus squirting out a cloud of ink.

"When the Sepia is frightened and in terror, it produces this blackness and muddiness in the water, as it were, a shield held in front of the body."

Aristotle, *Historia Animalium*

as dopamine, the neurotransmitter that is also found in the human brain, where it is plays an important role for both physical and mental functions, especially movement, learning, and memory. To date, there is no clear explanation of how it works in cephalopod ink.

Ejecting a squirt of ink is an enormously effective defence mechanism, which can be used as a sort of smoke screen when a predator is nearby. By varying the mucus content of the ejected liquid, a cephalopod can distract a potential predator by creating a pseudomorph that resembles its own shape or thin streams that look like the stinging tentacles of jellyfish. These keep their shape for a period of time in the water, serving as a decoy while the cephalopod jet propels itself to safety.

■ A jar of *Sepia* ink and 'black' rice and pasta that have been dyed with it.

Innards

The innards—liver, ink sac, glands, gills, hearts, reproductive organs, stomach, and intestines—are all enclosed in the mantle. In molluscs, as well as in fish, the liver and the pancreas are actually combined into one glandular organ, the hepatopancreas. But for the sake of simplicity, it will usually be referred to just as the liver. It secretes vital digestive enzymes and is also responsible for the exchange of nutrients and waste products that are later expelled via the anus and for the flow of water out through the siphon.

The digestive process is both fast and effective, which is important given that cephalopods often have a large intake of food over a very short period of time. As they have no enzymes capable of breaking down carbohydrates, they cannot eat plants and seaweeds and their diet consists solely of living creatures.

In the females of some of the species with ten appendages, for example *Loligo* and *Sepia*, there is a pair of bottle-shaped glands, known as nidamental glands. They are large enough to eat, either as part of the whole animal or cooked on their own. They secrete a gelatinous substance that helps to clump together the eggs into a mass before they are released.

Speed and cunning. Charles Darwin was a keen observer of the cuttlefish that he found in pools of water left along the shoreline by the retiring tide. They were hard to catch because they could escape into very narrow crevices where they attached themselves firmly by means of their suckers. "At other times they darted tail first, with the rapidity of an arrow, from one side of the pool to the other, at the same instant discolouring the water with a dark chestnut-brown ink." Charles Darwin, *The Voyage of the Beagle – Day 3 of 164* (1839).

◘ A squid (*Loligo forbesii*) that is split open along the back, exposing the entrails, including the liver, ink sac, and gills.

A Master of Disguises

Many species of marine animals benefit from a method of camouflage called countershading, an effect produced by having a light-coloured underbelly and a dark back. This helps to make them less visible, allowing them both to escape predators and to sneak up on prey. Some animals are also able to change the colour of their skin and their outward appearance, although some can do so much more quickly than others.

A colourful display of *Sepia* skin tones.

In the course of millions of years, in order to compensate for their lack of an outer, protective shell, cephalopods have developed an extremely sophisticated array of techniques for disguising themselves, which far surpass those of most other creatures. It is so successful that it sometimes makes it difficult to identify the different species accurately.

Most cephalopods can change the colour and even the texture of their skin. This ability varies from one species to another, but cuttlefish, especially those that live near the surface, are generally considered to have mastered the art to a greater extent than squid. The skin of cuttlefish can be compared to a large screen on which there is a dynamic display of colours and the patterns that they form. Some kinds of octopuses are also good at changing colour to blend in with their surroundings, while others have a more limited palette. In deeper waters, these displays are mostly limited to changing the skin tone toward reddish colours, so that the octopuses are less noticeable in the dark environment.

This fantastic masquerade is made possible by a series of small organs, known as chromatophores, iridophores, and leucophores, which are located in layers under the surface of cephalopod's skin, as well as a special muscle structure that drives it all. The precise arrangement of these organs varies from one species to another and for a given individual it can even differ from one part of the skin to the next. In principle, the skin tones come about in two different ways. One results in 'true' colours, which are due to special pigments. The others could be called 'structural' colours, which are created by the way light is reflected and refracted by the skin.

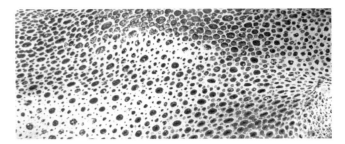

□ The chromatophores on the skin of a squid (*Loligo forbesii*) appear as small reddish spots.

Changing colour patterns even while being eaten. In 2016 one of us (Ole) had a unique experience at a sushi bar in Fukuoka, Japan while eating a still live Pacific flying squid (*Todarodes pacificus*). Its mantle had been cut up for *sashimi* and then placed again on top of the squid. Even while the chopsticks were picking up these pieces, their skin colour changed, forming a glowing pattern. This reaction is often an indication that the creature is feeling threatened—small wonder!

The chromatophores lie just below the surface of the skin. They are small, complex organs that are less that one millimetre in diameter and that each have fifteen to twenty-five radial muscles that are controlled directly by the brain. At their centre there is a small, elastic sac that contains a pigment, which in the case of octopuses can be either red, yellow, or brown and in the case of the large species of *Sepia* either red, yellow, or blackish-brown. These colours are particularly well-suited to provide camouflage at the depths at which these cephalopods live. When the muscle structures are activated, these pigments can be made to work together to create many other colours and distinctive patterns. When the muscles contract, the sac is stretched to form a larger, brighter patch of colour at precisely that place just under the skin. This happens incredibly quickly because the effect, unlike that in other animals such as amphibians, is not dependent on a hormonal signal. An octopus can change the colour of its entire body in a fraction of a second. The skin of cephalopods can contain several hundred thousand chromatophores, with octopuses generally having more than cuttlefish, which have more than squid. Although octopuses have more chromatophores, they are not able to activate them as quickly and imaginatively as cuttlefish. Apart from the camouflage provided by the colours and patterns themselves, cephalopods can also arrange them to create special effects such as large eye spots that may frighten an attacking predator.

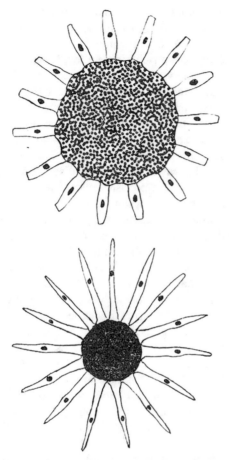

◘ Chromatophores are small, elastic sacs containing pigment that are attached to a set of radial muscles. When these relax, the pigmented area is expanded, making the colour less visible and changing the appearance of the skin.

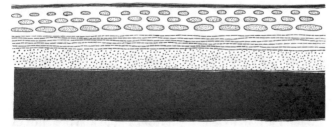

◘ The skin structure of cephalopods, which enables them to change colour, display iridescence, and reflect light. From the top: epidermis, chromatophores, iridophores, leucophores, and muscle layer.

When the chromatophore sacs are contracted, the colour effects are determined by the iridophores and leucophores that lie in layers beneath them.

The skin of an octopus turns blueish-red when it is blanched and cooked because the chromatophores, which contain red, yellow, and brown pigments, break up and the pigments seep out into the tissue.

The second colour-changing effect is created by iridophores, which are very thin, transparent crystalline chitin cells that are stacked in sheet-like fashion immediately under the chromatophores. By reflecting and refracting light, they produce a metallic, silvery, or violet sheen, as well as the green and blue hues that are lacking in the chromatophores. This is known as 'structural interference' and is similar to the way a soap bubble displays an iridescent colour spectrum. The precise mechanism controlling the iridophores remains a mystery although it seems that they are not in direct contact with the brain, but are controlled by chemical signals and consequently respond more slowly. What is special about the iridophores, however, is that they can also reflect polarized light. It would appear that cuttlefish can see polarized light and take advantage of this ability to orient themselves in relation to their environment, but it is not known whether cephalopods make use of polarized light as a communications tool.

What is the meaning of all these colour changes? This is an intriguing question to which researchers have not yet found the answers. Are they ways to communicate, or a means of camouflage, or simply some visual imagery? Many cephalopods display a particular, recurring pattern that is called 'the passing cloud display,' which resembles dark clouds passing over their body when other animals or members of their own species are nearby. Does it mean 'watch out and move over' or 'stop and pay attention to me?'

A third layer next to the muscle tissues contains leucophores, static flat white cells that do not change shape or appearance. They are unable to manipulate light or its colours and their sole functions are to provide a chalk white background and to reflect light, thereby reinforcing the effects created by the chromatophores and the iridophores.

As already mentioned, cephalopods are colour blind and, sad to say, they miss out on even their own fantastically colourful display!

In addition to colours and patterns, cephalopods have other tools at their disposal. Due to the three-dimensional nature of their muscle structure and because their bodies are muscular hydrostats, cephalopods can not only change overall shape but also vary the texture and corrugation of their surface. For example, octopuses and cuttlefish can contract their muscles to make their normally smooth surface look wavy or to form millimetre-sized protrusions, called papillae, which can be bumpy or spiky. This is especially useful for changing the appearance of the area around the eyes to hide them from a potential predator or to alter the shape of the head, so that it no longer resembles a living animal.

◘ An octopus (*Octopus vulgaris*) that has turned reddish-purple when its pigments seeped out into the tissue during cooking.

The way in which cephalopods are able to disguise themselves and the myriad appearances they can take on are not only a means to blend in with their surroundings on the seafloor. In other cases, they might resort to what is called 'disruptive colouration,' which involves taking on strongly contrasting patterns as a mechanism to break up the normal animal contours that would otherwise give them away to predators. This is well known to us from its use by the military to camouflage vehicles, ships, and personnel.

Not Just Colour, but Also Light

Many of the ways in which cephalopods conceal themselves are dependent on an interaction with passive light, whether reflecting it or refracting it. But more than half of the squid and cuttlefish and a few of the octopuses are also able to emit active light, an effect known as bioluminescence. The light is produced when certain chemical reactions take place in bacteria that live in symbiosis with the cephalopods inside special photo-active cells (photocytes) or organs (photophores) that are embedded in their tissue. It is possible for the animal to control both the colour and the intensity of the light.

A certain cephalopod (*Euprymna scolopes*) plays host to a species of bioluminescent bacteria called *Vibrio fischeri*. As these are not present in the newly hatched animal, the bacteria found in the surrounding seawater first have

Yet another way to escape detection. One cephalopod, the mimic octopus (*Thaumoctopus mimicus*), not only takes on a striped pattern that might frighten a predator but also uses its flexible muscle apparatus to make drastic changes in the way it moves so that it resembles a completely different animal. It is reputed to be able to mimic up to fifteen other species, most of which, such as the lion fish and the sea snake, are poisonous.

to establish themselves in its photophores. They develop together to form a single, integrated light-emitting organ.

Bioluminescence can serve a number of functions. It can act as a camouflage so that enemies are led astray because the cephalopod blends with the light surface of the water above it when it is seen from below. It can also cancel out dark shadows. Conversely, in the dark depths of the ocean, the light can act as a beacon to attract or locate prey, to scare off predators, to communicate with other cephalopods, or possibly to find a mate. For example, underwater observations of both a colossal squid, *Mesonychoteuthis hamiltoni*, and a giant squid, *Taningia danae*, have recorded frequent discharges of bioluminescent light flashes. Researchers think that this might be a mechanism to confuse their prey before attacking it or to measure how far away it is..

Sensory Systems in the Skin

Apart from its role in protecting the body of a cephalopod and as an organ that is able to conjure up all manner of disguises, the skin of a cephalopod also has specialized receptors that help it to see, taste, and smell.

Researchers have recently discovered that some cephalopods have specialized proteins, referred to as opsins, in their skin. Unlike the eyes of cephalopods, which are colour blind, these photoreceptors respond to different colours of light. So it is possible that this helps to compensate for that deficiency by allowing them to 'see' colours with the skin, perhaps enabling them to decide on the type of camouflage that would work best in a particular environment.

As already mentioned, their suckers are equipped with chemoreceptors that can pick up the 'taste' of the objects they are exploring by touch or to which they are attaching themselves.

Cephalopods have an olfactory organ, which is located behind the optic lobe. It functions as a sort of nose that can detect odours from molecules in the salt water that comes into the mantle. Among other functions, it can help to seek out potential prey and it can smell ink that has been spurted out by other cephalopods, which might be a sign of danger.

Cephalopods in Naples

Università degli Studi di Napoli Federico II, which was established in 1224, is the oldest public non-sectarian university in the world. Its old main buildings, which house some of the faculties, as well as its zoological museum and collections, are located right in the centre of the historical part of Naples. The department of biology is situated on the newer campus at Saint Angelo, outside of town. It is here that one of us (Ole) had the chance to visit Professor Anna di Cosmo, a zoologist who is also a cephalopod researcher and director of the museum.

When Anna described her work with octopuses, her deep love of, and great respect for, these peculiar animals were very evident. At the time, she was trying to discover, on a molecular and genetic level, what governs the ability of octopuses to adapt so quickly to changes in their environment. She and her colleague, Professor Gianluca Polese, have studied the interplay between the endocrine and nervous systems (the so-called neuroendocrine network) and the immune system in octopuses. They have identified the genes that underlie what is known as an epigenetic mechanism. It is possible that, as a function of the structure of the nervous system, this mechanism might constitute the basis for the peculiar adaptability on which the octopuses have depended in order to evolve successfully. It may also provide an explanation for the growth in the world's population of octopuses. Anna's research in this area has been made possible by the first complete sequencing of the *Octopus* genome in 2015.

Anna emphasised, time and again, that octopuses are a protected species, which is why she and her colleagues were continuing to work on formulating international rules governing animal welfare related to studies of octopuses and their forms of life. In this connection, she was very engaged in developing courses for the students at the university that would make the link between the purely zoological aspects of marine organisms and an understanding of food chains in the oceans and how they can be exploited more sustainably, among other ends, for human consumption.

Anna remarked that, as a zoologist, she found it difficult to understand people who will not eat animals. Both she and Gianluca, who had just dropped by her office, think that it shows respect for their research animals to eat them and not just discard them when the experiments have run their course. They told me about their studies to devise models for anesthetizing octopuses by dissolving particular anesthetics in the water in which they live. These models work for octopuses only because they live mostly on the seabed and move about very little. The models cannot be applied to studies of cuttlefish and squid, which are very difficult to keep in a controlled environment in a laboratory as they need to be able to move constantly in flowing water.

When I asked how best to see fish markets where cephalopods are sold in Naples, I was advised to visit the old market Porta Nolana. But I was also told that at a beach outside the city there is a dock for small fishing boats where fishers arrive with their catch in the morning and sell their wares directly to the public. This is where Anna and Gianluca themselves buy some of their research animals, often in competition with researchers from the Stazione Zoologica Anton Dohrn, a place that I was to visit later. We joked that these fishers are the scientists' 'octopushers.'

When I prepared to return to town, Gianluca offered to drive me there as he was heading back to the old part of the university. Gianluca is also a zoologist with a particular interest in chemoreception, especially the sense of smell in animals. When I asked him which animals he studies, he replied somewhat humorously, channeling our earlier conversation about cephalopods, "all animals that can be eaten." This discussion took place as I sat on the back of his motorcycle while he made his way through the unbelievably dense and chaotic Neapolitan traffic.

The next day I visited the Porta Nolana market to see how fish, shellfish, and cephalopods are sold in the open in the narrow alleyways of the old part of town. Here one finds a variety of species of large and small octopuses, some still alive, as well as large cuttlefish and a whole range of different squid. Apart from cephalopods, there is an abundance of fresh fish and shellfish.

Naples is renowned internationally as a centre for marine biology research carried out not only at the university but also at a specialized institute, Stazione Zoologica Anton Dohrn, named after its founder who established it in 1872. Initially it was set up as a private establishment, but it was turned into a national institute under the aegis of the government in 1982. It was first housed in a distinctive building opened in 1874 in a delightful seaside park called Villa Comunale. Among Dohrn's early supporters were a number of distinguished biologists, including Charles Darwin, who also contributed financially. Already in the 1880's the institute became famous around the world thanks to the work of Salvatore Lo Bianco, who developed sophisticated methods for preserving and storing marine specimens.

Since its founding, Stazione Zoologica Anton Dohrn (NZS) has been the workplace for a long succession of researchers and during a significant portion of the past century was the pre-eminent site for the study of octopuses. Over time, its original focus on physiology and morphology shifted to investigations into the nervous systems, brain, learning, and behaviour of these creatures. This change mirrors the broader scientific trend to integrate zoology with the theory of evolution.

NZS is housed in an imposing building with a large inner courtyard also housing the entrance to the publicly accessible aquarium. Until recently, the aquarium still had the more than one hundred and forty-year-old tanks with glass panes. To this day one can see that Anton Dohrn was inspired by Goethe's ideas about the unity of art and science, which is reflected in the impressive frescoes found in the old, well-preserved library. At the time of my visit, the whole building was undergoing extensive renovations, partly because the old aquarium tanks were not up to modern standards. They were not insulated and, as a consequence, water seeped out, which over time damaged the brick work and the foundations.

I was able to obtain an appointment to speak with the head of the department of biology and the evolution of marine organisms, Dr. Graziano Fiorito. He is world-famous for his work on the behaviour and neuropsychological conditions of octopuses, which has been his main research interest since 1983. In 1992 his findings shook the international scientific community when he published an article in *Science* in which he stated that octopuses are able to learn by observation and then put this knowledge into practice. He concluded that this ability showed that octopuses can integrate information to create adaptive behavioural models. This assertion transcends some generally accepted biological theories as it indicates that an invertebrate like an octopus has evolved along the same principles as vertebrates, that is, like ourselves. Even though this experiment could not be reproduced by others, his article led to completely new perspectives from which to view these creatures.

Dr. Fiorito met me at the reception area of NZS and apologetically told me that he had a very busy day with guests, meetings with students, and a visit from a television crew. The latter were there to record a documentary on how an understanding of the intelligence and behaviour of octopuses can be used to develop more biologically inspired intelligent robots. In order to find some time to talk in a quieter atmosphere, he invited me to have lunch at a small, nearby restaurant. On the way, he told me that there is a special relationship between this restaurant, which started life as a wine bar, and NZS, as they were established at about the same time. Six generations later, it was still run by the same family and it had become a meeting place for NZS researchers and their guests to talk science over a meal. Following along in that tradition, we were greeted warmly by the host who commented that we were some of the 'aquarium people.'

I quickly realized that Dr. Fiorito does not share all of Anna di Cosmo's perspectives on octopuses. He said that since he began to regard their arms as being like fingers that reach out to him, he has been unable to eat octopuses. He has as much respect for the animals as does Anna, but that does not lead him to think that they should be eaten at the conclusion of an experiment following which they are to be put to death. He put it this way—by interacting with him in the experiment, the animal has given him something and that implies that there is an implicit contract not to take anything from it. Further to his outlook on respect, he considers it of utmost importance that he and his fellow researchers should never give these animals a name and that they should in no way establish any personal connections to them, as one might with a pet animal. The octopuses are given a number and regarded as experimental subjects and no more.

Dr. Fiorito is very concerned about animal welfare. In particular he has been involved with implementing regulations for the European Commission with regard to the treatment of cephalopods not only as research animals, but also in connection with how they are caught, transported, and killed for food. It seemed to me that his rather forceful and very definite position on these matters was grounded in his knowledge of octopuses and their very highly evolved brain and nervous system.

He asked me whether I had noticed the wide channel in the middle of the octopus arms in my lunch salad. I had and also knew that in the middle of each arm there is a thick cable, which is a bundle of some 30,000 nerve cells. It is similar to a notochord in vertebrates. Dr. Fiorito is of the opinion that the arms of an octopus are much more independent with regard to the nervous system than was previously thought.

After returning to NZS, I was able to see the research laboratories and meet some of Dr. Fiorito's colleagues and students. One of the junior researchers, Pamela Imperadore, had chosen to work on a project involving the octopus nervous system for her Ph.D. She told me about her efforts to understand how octopuses regenerate damaged brain nerves. Colour changes in an octopus are controlled by the central brain, which can communicate with the muscles that control the contraction of the chromophores distributed in the skin. Much to her surprise she had discovered that brain damage immediately causes the octopus skin to turn white. But the ability to create a red colour returns very quickly, in a matter of a few weeks, even though it takes several months to regenerate the damaged nerve connections in the brain that were the original cause of the animal's inability to activate the chromophores. This amazing ability to establish new nerve connections in a decentralized manner has caught the attention of neurologists who are concerned with the regeneration of nerve cells in humans who have suffered a spinal cord injury.

Pamela took me to the room in the basement that is equipped with tanks for the research octopuses. The tanks all have a glass panel, which allows the animals and the humans to look at each other. Each octopus has a small den to which it can retreat, for example, when it is eating. The room itself actually has a cave-like feel, with special lighting and curtains that can be drawn between the rows of tanks. Each tank is separated from the one next to it with a glass window and a screen. The screen can be removed so that the octopuses in neighbouring tanks can see each other when the researchers are trying to find out whether they can learn by observation. Can the octopus in one tank acquire a skill, for example, opening a closed box or unscrewing the cap on a glass jar, simply by seeing how the octopus in the other one does it?

Pamela also showed me a number of black and transparent boxes with lids and drawers in which the researchers placed food to study whether the octopuses, under varying circumstances, could open the boxes and get to the food. There were also some special transparent systems of pipes in which a live crab, the favorite food of the octopuses, could be placed. There were three different ways to open them up, either via a lid with hinges, a simple lid, or a screwed on lid. Pamela also showed me some sticks of different colours that, when touched slightly, can be used to stimulate the animals in a positive way to investigate their ability to learn, remember, and particularly to link a colour to an event. In contrast to earlier experiments at NZS in which the octopuses were stimulated negatively with electric shocks, only positive stimuli are now used.

When we returned to Dr. Fiorito's office at the conclusion of our discussion, I could see that, like Anton Dohrn, he is aware of the importance of art and esthetics in relation to scientific endeavors. On the walls there were pictures of drawings of different cephalopods made by Japanese artists and small sculptures and artistic replicas of cephalopods are found everywhere on desks and shelves. And as a goodbye present he gave me a very heavy, very large book on cephalopod patterns, which he had co-authored. He had employed an illustrator for a period of three years to create meticulously detailed drawings of an assortment of cephalopods and of their anatomical structures.

When I bade him farewell, Dr. Fiorito was on his way, together with his students, to meet up with the television crew down in the 'octopus cave' in the basement. He expressed his concern about what the television crew might hope to accomplish with their visit and said that "they want to regard the animals as mere machines, but that is not what they are." I replied that possibly seeing them in real life would convince them that the octopuses are indeed not machines.

How Cephalopods Are Caught

Archaeological evidence indicates that humans have been catching cephalopods for food for at least 4,000 years. Since ancient times around the Mediterranean, octopuses have been trapped in clay pots fastened to a rope and lowered to the seabed. The Romans learned the technique from the Greeks, who may have copied it from the Egyptians. It is likely that Japanese fishers have been catching them in this way for an equally long time. This method, which employs both pots and bottles with a not too narrow neck, is the most effective way to catch octopuses and is still in use.

Depending on the scale of the fisheries and the geographical location, octopuses are presently caught in a whole range of ways: by trawling, in traps (pots, bottles, cylindrical nets, and boxes with lids), with baited hooks, using spears and harpoons, and simply with bare hands. There are, unfortunately, some very real problems associated with two of these techniques. Bottom trawling does collateral damage to the sea floor itself, thereby disturbing other sea life, and it tends to scoop up a large proportion of bycatch, with octopuses making up less than half of the total catch. As most cephalopods can move very quickly using jet propulsion, specialty trawling equipment is required to catch the biggest specimens. The traditional method using pots risks destroying the eggs and hatchlings that are being sheltered in them. Given that these creatures mate only once and then die, a whole new generation might be lost. And the bait also attracts other marine life, which is not always desirable. The most commonly caught species on a world-wide basis is *Octopus vulgaris*, which is sold either fresh or frozen.

Squid are often caught using methods that have less of an impact on finfish stocks and that work by activating a naturally occurring food chain. Lights are used at night to attract plankton, which are phototropic, toward the fishing boats. This, in turn, lures the small fish species that feed on them and the cephalopods that prey on the fish toward the surface. In California, for example, the lights entice various species of *Loligo* to an area where they can then be trapped in a seine net. By far the two most important commercial squid fisheries are for Japanese flying squid (*Todarodes pacificus*) in Asian waters and the Argentine shortfin squid (*Illex argentinus*) in the Atlantic off the southeast coast of South America. Here, most of the squid are caught using metal lures and blinking, baited hooks attached to jigging lines. There is much less waste and the method yields the highest quality of squid, which can be sold for the best prices. It is estimated that Japanese flying squid make up about a half of the total global cephalopod catch. The largest market for *Todarodes pacificus* is in Japan, where about 70 percent of the catch is sold fresh to make *sashimi*. The rest is frozen or used to make dried, salted, and smoked products. The innards can be used in fermented products.

Seen from space. The squid fishery for *Todarodes pacificus* is on such a large scale that photographs taken by astronauts on the International Space Station in December 2017 clearly show the glow of green LED lights from the hundreds of fishing boats in the Gulf of Thailand and the Andaman Sea.

◘ Squid fishing with lights at night in Thailand.

Like octopuses, cuttlefish have been a source of food for humans for centuries. Cuttlefish live close to the seabed near the shorelines and some species can grow to a size of about sixty centimetres. They are caught by bottom trawling and in seine nets, using blinking jigging hooks, and in submerged pots. In Europe, *Sepia officinalis* is one of the most important commercial species and is usually harvested by targeted trawling or as bycatch when fishing for bottom dwelling fish. Cuttlefish are marketed fresh or frozen. China, followed by Thailand, are the primary sources of cuttlefish.

Cephalopod Fisheries around the World

Squid, octopus, and cuttlefish are caught in all the oceans of the world, both on a large, virtually industrial scale using factory ships that scour international waters and by people living in small, artisanal fishing communities. Cephalopod fisheries have doubled in size since 1980, but they are still dwarfed by the total annual harvest of all marine creatures, both wild and farmed, and account for a little less than 5 percent of the overall volume. Nevertheless, together with tuna, shrimp, and lobster, they are regarded as one of the most valuable marine resources. Most of the tonnage is destined for human consumption, with only a small portion used as bait for other fisheries. Almost all cephalopods are caught in the wild, as raising them in aquaculture is still a very limited and somewhat experimental undertaking, although major efforts are underway expand its scope.

Despite the economic impact and food security importance of the cephalopod fisheries, it is very difficult to obtain reliable statistics about their extent because international waters are not subject to quotas, not all countries report their catch, and surveys generally include only the quantities taken by commercial enterprises, without accounting for small-scale unregulated local activities. In some cases, a significant proportion of the catch is lumped together as a group that is undifferentiated by species. And because cephalopods are often traded actively and imported and exported several times over it is not easy to trace their provenance and to enforce quality control. For example, octopus used for sushi can come from a third country via Japan. All of these factors further complicate attempts to determine the extent to which the cephalopod stocks are being managed in a sustainable fashion and whether the various populations are increasing, stable, or declining, especially in the face of possible over-fishing and the effects of climate change.

Squid, which are found primarily in temperate and tropical waters of all the oceans, are currently estimated to account for 80 percent of the total cephalopod tonnage. Of the 290 species of squid, only about thirty to forty are of commercial importance, based mainly on whether they can be caught in significant numbers near the surface, are of sufficiently large size, and have an appealing taste and texture. The most important species belong to the flying squid family (Ommastrephidae), with *Dosidicus* spp., *Todarodes* spp., and *Illex* spp. most predominant, and to the pencil squid family (Loligonidae). Thailand, Spain, China, Argentina, and Peru are the world's major exporters of squid and also of cuttlefish.

Octopuses and cuttlefish each make up about 10 percent of the balance of cephalopod landings. Octopus fisheries have been growing in importance over the past three decades and the volume has been increasing steadily, almost doubling during that time. Worldwide, fishers from some ninety countries are involved in the harvest of more than twenty different species, with *Octopus vulgaris* and *Octopus maya* being two of the more important ones. At one time, the octopus fisheries in Spain and Portugal accounted for a large share of the catch, but the total European landings have decreased by more than half in the past thirty years. Atlantic octopus fishing has moved south to the coast off northwest Africa, where there is an ideal marine environment due to the nutrient-rich water caused by the upwelling of the ocean near the continental shelf. Other important octopus fisheries are located in waters near China and around the Yucatán peninsula in Mexico.

◧ Squid being sun dried in Thailand.

▣ Squid (*Loligo forbesii*) caught in the North Sea on a long line by an angler.

Tagging along on a Fishing Expedition off the Coast of Granada

Motril on the coast of Granada is the sole remaining fishing port in the province and further development of this economic sector has come to a virtual standstill in the area. There are now only 120 fishers working on about thirty boats, of which twelve are trawlers. Nevertheless, thanks to contacts in the city hall of Motril and the local fishers' association, Cofradía de Pescadores de Motril, it had been possible to arrange for Klavs and Ole to go out on a small trawler for a day of fishing along the shoreline. Ignacío López Cabrera, the captain of the boat, who is also president of the Cofradía, had promised to take us along with his crew of six to observe their work.

Ignacío makes a day-to-day decision as to what to aim for in the catch, based on its market value At that time, toward the end of September, it was not yet the peak of the season for catching octopuses, which falls in the last two months of the year. So the boat was not setting out specifically to harvest cephalopods, but we were advised to keep a lookout for them anyway, as it was likely that some specimens would be pulled up in the net. The captain had decided on fishing in shallow waters, at a depth of ninety to one hundred metres, in parallel to the coastline going toward Almuñecár. There would be three trawls in all, each lasting three hours, meaning that we would not return to port until six o'clock in the evening. We cast off at half past five, with a very long workday ahead for the men.

On the way out, we met the 'night shift'—the fishing boats that were returning with what the Spaniards call *pescados azules* (blue fish), in this case mostly sardines, anchovies, and mackerel, which hunt in schools. They are caught in the dark, often with the help of lights, and sold at auction in the morning.

Another passenger on the trawler was Inmaculada, (Inma) Carrasco Rosada, a biologist. She had been working with the fishers' association with the goal of improving their approach to fishing and obtaining a better economic return for their products. She had come along to guide us through the day's events and also to serve as a very helpful translator.

Our expectations were a mile high because we had never tried anything like this before. It turned out that we were not disappointed and the catch did include some octopuses. But the downside was that we witnessed two completely different aspects of this type of fishing. It demonstrated the extent to which the Mediterranean has become a dumping ground for plastic waste and also that trawling is not the right way to exploit marine resources.

On the outward journey, Ignacío and Inma told us about the major challenges facing the fishing industry in the Mediterranean. Ignacío was well aware that things have to change and that is why he and his association had started working with Inma to look for new and more sustainable ways to fish. They were both very active in trying to draw attention to the environmental problems of the Mediterranean, the most polluted sea in the world. But the everyday reality was that there were seven families that depended on this trawler for their livelihood. And the market prices were so low that it was difficult to set aside money to invest in new fishing methods.

After a while, dawn broke and we could see the beautiful Granada coast as we sailed along on the calm, blue waters. It was time to drop the trawl net for the first pass. It is dangerous work as the six men had to coordinate between the winch to unspool the net and the speed of the drum, while attaching the trawl doors that hold the net open and the warp lines that pass through a block to keep it in place behind the boat. While all this was taking place, the captain had to control the speed of the boat so that everything went smoothly. Then it was simply a question of waiting while the trawler sailed very slowly at a speed of five to six kilometres per hour. Some of the crew disappeared into the cabin beneath the deck to catch a little sleep. Others made coffee in the galley and sat down to read the newspaper.

When it was time to haul in the net, the landlubbers were placed in the corners of the boat, safely away from the winch and the warps attached to the trawl net, while the boat was stationary in the water. In no time at all, as the trawl was coming onto the boat, large flocks of seagulls appeared. They knew what was in store. And just before the last part of the net was hoisted out of the water, other 'guests' were spotted near the boat—dolphins that also had hopes of getting something good to eat. Everyone was waiting with bated breath to find out what had been caught—not only the visitors who had never before seen this type of operation, but also the fishers who depended on getting a good catch of fish that could be sold at a decent price. At one time, they had been paid according to what was caught, but now they were on fixed wages. Nevertheless, in the long run if the catch is poor and there is no money to be made, the boat will eventually be taken out of the water.

As the farthest end of the trawl was coming onto the boat, it was lifted up with a hook so that it was suspended freely over the deck before the rope that closed its bottom was loosened and the catch fell out. When this happened we were scared out of our wits by the unforgettable sight before us. The 'catch' could best, if somewhat vulgarly, be described as a large, muddy pile of shit, made up of all kinds of plastic garbage—small and large bits and pieces, ropes and netting, tubes, bottles, crates, etc.—mixed up with writhing and half-dead fish, starfish, sea urchins, and small crustaceans. In addition there were lots of weird brown, slightly spiny cylindrical objects, which were ten to twenty centimetres long and up to five centimetres in diameter. They were sea cucumbers (*Stichopus regalis*). As this disgusting heap spread out on the deck, we could spot, here and there, the arms of octopuses that were trying to crawl out the pile. So we were in luck after all.

What happened next demonstrated that ocean fishing in this way is still just a version of hunting and gathering. The six crew members got down on their knees and started to dig around in the mess with their bare hands, helped only by some small, worn sticks. The valuable parts of the catch were picked up one at a time and thrown into baskets that were placed next to them, to be sorted more carefully later. The plastic garbage was removed and thrown onto an ever-growing pile. Everything that did not make the first cut, including undersized fish, was pushed out through a slot in the deck and sent back into the sea, where the seagulls and dolphins were waiting for their meal. We were surprised to see how much went overboard. We were especially puzzled that masses of small, shiny fish that looked as if they would be really delicious were discarded as well. The fishers said that they had no market value. As most of these fish were dead before they were thrown back in the water, this seemed like a huge waste.

We were naturally interested in seeing which cephalopods had been caught. There were a few squid and cuttlefish, which was quite coincidental because they are normally caught by being lured closer to the surface or by using other types of nets. There were several species of octopuses, including *Eledone cirrhosa* and *Octopus vulgaris*. As the latter are the only ones that can be sold for a decent price at the auction, they were separated out and thrown into buckets that were covered with a net that had a hole in the middle. It acts as a trap because if they try to crawl out over the edge their arms become hopelessly entangled in the mesh. But they nevertheless shot out their long arms and used their suckers to 'taste' their new, foreign environment. Undersized specimens and those that had no commercial value because consumers do not like their taste were thrown overboard. When they hit the surface of the water they sent out a cloud of ink to act as camouflage. We could see that the cloud dispersed only slowly, giving the octopuses enough time to jet propel themselves to safety near the seabed.

Once things were somewhat under control, the trawl net was lowered once again. The crew then turned their attention to dealing more fully with the initial catch. They needed to sort the fish by type and tidy up the deck, which was still littered with mud, crustaceans, and a lot of small fish. The baskets with the different fish were picked over once again and hosed down. The cleaned fish were packed in boxes and carried down via a small ladder into the cool room in the hold. At that point the fishers started in on the only production aspect of the work that takes place on this type of trawler, namely to cut up the sea cucumbers.

A sea cucumber is basically an elongated cartilaginous capsule with holes in each end that acts as a mouth and an anus, respectively. These are connected by a looping intestine with the rest of the space being taken up by other organs and sea water. The animal exists by taking in nutrition from the mud on the seabed and passing it through its digestive tract. Whole sea cucumbers are eaten in some parts of Spain, but the fishers of Motril are interested only in harvesting the intestines. These are sent to Barcelona, where they are considered to be a great delicacy, and consequently can be sold at a premium. So the crew got to work. They sat hunched over on boxes, balanced a wooden plank on a couple of buckets to act as a cutting board, and used their pocket knives to slice off the ends of the sea cucumbers, slit the bodies open, and tear out the intestines. Some of the intestines disappeared into the galley and we were soon presented with some delicious little snacks—pieces of intestine sautéed in a little olive oil with garlic. They were exceptionally tasty and had an interesting crunchy texture.

◘ Hauling in the trawl net on a fishing boat in the Mediterranean.

In the meanwhile, two of the fishers were preparing lunch on a gas cooker in the galley, partly from ingredients caught in the morning. It consisted of rice, fried chicken, grilled weever fish and small crustaceans, and a fish soup, all of it served with some crusty bread. One of the fishers even carefully whisked together some mayonnaise in an old mortar. When the food was ready we all crowded around the small table in the galley, although with guests on board there were not enough chairs and some had to stand. In the absence of plates, the food was spread out on newspapers and we took turns using the available spoons to eat the excellent fish soup. It was all washed down with water and soft drinks.

Once we had finished eating it was time to haul in the trawl net a second time, a repeat performance of the morning's effort. The catch was somewhat bigger, but the sight of the muddy heap that looked like garbage was just as appalling. By the time we saw it a third time later

on we decided that we had had enough of fishery of this type. But we were still fascinated by the octopuses that were trying to writhe their way out of the heaps. The total octopus catch for the day included some large specimens and amounted to 20 kilograms, considerably less than the 120 kilograms or so that can be harvested on a good day at the height of the season in November and December.

With the third pass of the net complete, we started the journey back to the harbour in Motril to deliver the catch to the auction hall.

When we arrived back at the dock around six o'clock in the afternoon, a surprise was waiting for us. Instead of a truck to pick up and take away the fresh catch, there was a special truck to collect the plastic and other garbage that the fishers had gathered from the ocean that day. It was very discouraging to see how much of it there was. Its volume was actually greater than that of the fish and octopuses together. Apart from disposing of it, this garbage was also destined for a research project that is mapping the extent of plastic pollution in the Mediterranean. Much of what is gathered off the coast of Granada can be attributed to agriculture. One of the main components turns out to be the plastic covering the greenhouses in which a significant proportion of Europe's tomatoes are grown. In the past, the waste plastic was thrown into gullies that were washed out by torrents of water from the snow melt on the surrounding mountains, carrying the plastic with them into the sea.

Apart from the depressing sight of so much garbage being taken away, our return had a familiar aspect that we had experienced in other old-fashioned harbours, namely a local 'welcoming committee.' Both young and old gathered around the fishing boats to ask about the day's catch, to which captain Ignacío replied that it had been quite satisfactory.

As soon as we had tied up, the fishers loaded the fish boxes onto a pushcart and wheeled them straight into the auction hall. Within the hour most of it was sold and on its way to local restaurants and fishmongers. Some of the catch was wrapped in plastic bags and sold directly by the fishers to the local people. Anything that could not be sold at auction at the minimum price was donated to deserving, needy people. This is an arrangement that the Cofradía had initiated to ensure that nutritious food would not go to waste. Keeping this in mind, it was even more thought-provoking that over half of any given catch was dumped from the boat back into the ocean. These fish were either dead or ended up as food for sea gulls and dolphins.

On the way from the pier to the auction hall we passed some smaller boats that were equipped with large light projectors used at night to catch squid and cuttlefish in nets. In addition, there were piles of some strange-looking pots. They are called *alcatruzes* and are the most traditional way of trapping octopuses. Although they are no longer elegant ceramic amphorae but more utilitarian-looking and made of plastic, they are still lowered to the sea bed just as in ancient times. Each pot has a capacity of about two litres and contains no bait. The octopuses seek them out as safe shelters and make no attempts to crawl out of them when they are being hauled to the surface. Fishing using *alcatruzes* accounts for about half of the total octopus catch in the Mediterranean and others are harvested by using small net traps, *potera*. These two ways of catching octopuses are very seasonal and the pots and traps are set out only in November and December.

After the auction was finished it was time to relax with *tapas* and a glass of cold beer in the local 'hole in the wall' establishment. It was a hive of activity, both inside and on the outside where people were lined up along the wall, which had a shelf on which to place their food and drink. The *tapas*, made with fresh fish, shellfish, and octopuses, were prepared on a hot grill behind the bar counter. The deal was that as long as one continued to order beer, *tapas* were served on an 'all you can eat' model. And after a twelve hour day at sea, it was easy to tuck lots of them away.

Cephalopods Are Nutritious and Tasty, Too

The taste of raw cephalopods can be compared to that of their relatives in the other orders of molluscs and also partly with finfish, especially when it comes to umami. But the overall taste experience, particularly the mouthfeel, varies widely, for example, between that of octopuses and bivalves or between squid and finfish. This is due to the fundamental difference in how the muscles of the different species are built up. While the protein content of bivalves and octopuses is comparable, the vast difference in the texture of the muscle meat is a reflection of where they live and how they move about. In addition, there is considerable variation in the substances found in different species and this affects their nutrient contents.

Nutrients in Cephalopods

The available data base information for the nutrient composition of cephalopods is unfortunately not entirely precise as it reflects studies on a variety of species that were caught in different locations and at different stages of their life cycles. Nevertheless, the data bases can be used to construct a composite picture that gives a general impression of the potential value of different cephalopods as a food source. The table below provides comparative figures for the three main types of cephalopods—octopuses, squid, and cuttlefish—together with those for another mollusc (blue mussel), a lean fish (cod), an oily fish (salmon), and a terrestrial animal (chicken).

It is worth drawing attention to some of the more salient points that emerge from this comparison. Concerning macronutrients, the protein content of the cephalopods is very similar to that of the other animals but they contribute many fewer calories to the diet and their muscle meat is generally leaner. On the other hand, both squid and cuttlefish have a fairly high cholesterol count. Apart from octopuses, the fats are predominantly polyunsaturated fatty acids (EPA and DHA).

As can be seen from the chart, cephalopods are rich in essential minerals. For example, they are an especially good source of calcium, and iron and sodium are much more abundant in octopuses and cuttlefish than in fish and chicken. Recent studies undertaken on cephalopods caught in the Mediterranean have also found that they are able to make a meaningful contribution to the daily intake of micronutrients, in particular copper, selenium, zinc, and chromium.

With regard to vitamins, molluscs generally have very little vitamin K and no vitamin D, which is abundant in oily fish. Octopuses are, however, an excellent source of vitamin B_{12}, which is not readily available from some of the more commonly eaten meats and completely absent in plant-based food.

▫ Comparative table showing contents of water, calories, nutrients, minerals, and vitamins in octopuses, squid, cuttlefish, blue mussels, cod, salmon, and chicken

Per 100 grams	Octo-puses	Squid	Cut-tlefish	Blue mussels	Cod	Salmon	Chicken
Water (g)	82	78	81	81	73	65	73
Energy (kcal)	82	92	79	86	143	208	143
Protein (g)	15	16	16	12	17	20	17
Carbohydrate (g)	2	3	0.8	3.7	0	0	0.04
Dietary fibre	0	0	0	0	0	0	0
Sugar	0	0	0	0	0	0	0
Fat (total) (g)	1.0	1.4	0.7	2.2	0.7	13	8
Fatty acids							
Saturated (g)	0.2	0.4	0.1	0.4	0.1	3	2.3
Monounsaturated (g)	0.2	0.1	0.08	0.5	0.1	4	3.6
Polyunsaturated (g)	0.2	0.5	0.13	0.6	0.2	4	1.5
Cholesterol (mg)	48	233	112	28	43	55	86
Ca (µg)	53	32	90	26	16	9	6
Fe (µg)	5	0.7	6	4	0.4	0.3	0.8
Mg (µg)	30	33	30	34	32	27	21
P (µg)	186	221	387	197	203	240	178
K (µg)	350	246	354	320	413	363	522
Na (µg)	230	44	372	286	54	59	60
Zn (µg)	1.7	1.5	1.5	1.6	0.5	0.4	1.5
Vitamin C (mg)	5	5	5	8	1	4	0
Folate (µg)	16	5	16	42	7	26	1
Vitamin B_{12} (µg)	20	1.3	3	12	0.9	3	0.6
Vitamin A (IU)	150	33	375	160	40	193	53
Vitamin D (IU)	–	–	–	0	36	441	8
Vitamin E (mg)	1.2	1.2	–	0.6	0.64	3.5	0.27

Source: USDA (https://ndb.nal.usda.gov/ndb/)

How Do Cephalopods Taste?

Like the taste of fish, shellfish, seaweeds, and other fresh ingredients from the sea, that of cephalopods varies greatly from one species to another, but in all cases it is characterized by the presence of umami. This taste is attributable to their significant content of free amino acids, especially glutamic acid (and glutamate salt) and free nucleotides, particularly inosinate and adenylate. These substances are a source of umami and the synergy between them reinforces it. More than half of the free amino acids in cephalopods are of the type that we associate with the taste of seafood, namely arginine, glutamic acid, alanine, and glycine.

The large quantities of free nucleotides are due to the fact that molluscs in general, including cephalopods, have elevated levels of ATP (adenosine triphosphate), the substance that the muscles use to generate energy for movement. This has important implications for how cephalopods are caught. Stressing the muscles as little as possible before the animal is killed will lead to less ATP being used up, so that under the right conditions more of the free nucleotides will be present in the subsequent raw ingredients. Squid have an especially high adenylate content, up to 184 mg/100 g, which is as much as in scallops and six times as much as in sun-ripened tomatoes. Cephalopods contain a fair amount of free glutamate, up to 110 mg/100 g, somewhat less than in scallops (140 mg/100 g) but more than in chicken (22 mg/100 g).

Because the salt content of the cells of marine animals is lower than that of the surrounding sea water, these organisms need to compensate by increasing the concentration of substances that can counterbalance osmotic pressure. This is true both for fish and molluscs such as bivalves and cephalopods. Among the substances that help in this regard are a number of amino acids and trimethylamine N-oxide (TMAO). Bivalves have an especially high level of amino acids, including glycine, glutamic acid, alanine, proline, and arginine. In addition, glycogen, which fuels the movement of fast muscles, helps to elicit their sweet and spicy taste.

◘ Grilled long-finned squid (*Loligo forbesii*) served in the classical manner with lemon, flat-leaf parsley, and roasted garlic.

Cephalopods require more trimethylamine N-oxide than bivalves in order to maintain the osmotic balance with respect to the surrounding sea water, but they contain lower levels of substances that have a sweet taste, especially the free amino acids glycine and alanine. And as trimethylamine N-oxide is tasteless, cephalopods taste less sweet than bivalves. When cephalopods die, however, the trimethylamine N-oxide in their muscles is converted with the help of the animals' own very active enzymes to trimethylamine and they quickly start to give off an unpleasant 'fishy smell.' The reason why dead freshwater fish and crustaceans are less likely to give off these odours is that they contain less trimethylamine N-oxide (0.5 mg/kg) than salt water fish and cephalopods (40–120 mg/kg).

◘ The giant squid *Architeuthis* caught on a line baited with a smaller cephalopod.

Giant squid—no thanks! At first blush the giant squid, *Architeuthis*, would appear to be a good way to put a lot of food on the table. Unfortunately, it tastes and smells horrible because, like other cephalopods that live at great depths, it accumulates a large amount of ammonium chloride in its body. Ammonium chloride is lighter than sodium chloride (ordinary table salt) and, as a consequentce, helps the squid to remain buoyant despite the great pressure exerted by deep water.

Preferences with regard to the taste of prepared cephalopods are very closely linked to particular food cultures. Generally speaking, the muscle meat has a mild taste, has practically no aroma, and can easily blend in with other flavours. So when mixing it with other stronger tasting ingredients, care is needed to prevent the subtle taste of the cephalopods from getting 'lost' completely, with their texture as their only surviving characteristic in the dish. This dichotomy is illustrated by how cephalopods are often served in Asia. In Japanese cuisine the tendency is to try to preserve the natural taste of the raw ingredients themselves, possibly seasoned with a little marinade. In Vietnam, Thailand, and China, dishes with spicier, powerful, and fish-like tastes are more highly regarded.

Storing dead fish, crustaceans, and cephalopods at high temperatures and for long periods of time can lead to the formation of what are called biogenetic substances, which can be poisonous to humans. These substances result from the breakdown of the raw ingredients by their own enzymes and from microbial activity. The formation of certain biogenetic amines, for example, histamine derived from the amino acid histidine, poses a particular problem. Fortunately, the histidine content of cephalopods is very low.

Environmental Pollutants in Cephalopods

Contamination by the environmental pollutants and heavy metals found in oceans everywhere is a growing concern that affects all edible marine species to varying degrees. Among the worst of the problems is the accumulation of mercury, arsenic, and potassium in finfish and shellfish, cadmium in squid, and copper and zinc in shellfish and molluscs. While all cephalopods suffer from a similar fate, their relatively short life span ensures that the problem is less pronounced in them than in the longer-lived predator marine organisms that are higher up in the food web.

As cephalopods are becoming a more significant food source it is necessary to ascertain the extent to which they have accumulated environmental contaminants. On this basis it will then be possible to undertake a risk analysis of a human diet that includes a greater proportion of cephalopods. It is also important to differentiate between the levels of these substances in the muscle meat and in the innards. In the case of the latter, the readings are typically done on the hepatopancreas, the gland that secretes the digestive juices.

One study measured the content of biologically non-essential elements (mercury, cadmium, lead, and arsenic) in all three groups of cephalopods caught in the Mediterranean. It showed that octopuses have the greatest concentration of environmental toxins and squids the least. Apart from mercury and arsenic, which are evenly distributed in both the muscle meat and the innards, the hepatopancreas had the highest values for these substances.

The collective assessment of the health risks associated with eating cephalopods undertaken on the basis of the maximum weekly intake indicates that cadmium poses the biggest problem. One seventy gram portion can contain an amount of cadmium that is equal to 36 percent of the weekly maximum. It is, therefore, recommended that certain individuals, for example, pregnant women, should avoid eating the innards. Mercury and lead levels are so low that they do not pose a health risk for humans. The arsenic content is greatest in the various species of cuttlefish. But although the level is relatively high, it is not a cause for concern because in cephalopods it occurs in the organic variety, which is harmless.

Based on the integrated results of the studies, there is no overall reason to warn people against eating the meat from cephalopods.

Bacteria and parasites. Cephalopods live in the same waters as fish and have a similar diet. As a consequence, they are equally vulnerable to attack by bacteria and a variety of parasites, for example, the nematode anisakis. It is, therefore, of paramount importance to observe equally careful safe food handling practices for cephalopods as for fish.

◧ Boiled octopus arms sold at a Japanese market.

Buying, Preparing, and Storing Cephalopods

The idea of cooking with cephalopods can seem more than a little daunting to those who have never tried their hand at preparing them. But as we shall see, apart from cleaning them and dealing with the ink that might seep out of the innards, they are not really all that difficult to deal with. And it might be possible to avoid this part of the process altogether, if one is able to get them cleaned and cut up at the fish market or buy prepared ones in frozen form.

In culinary terms, there is a great deal of emphasis on the muscle mass, which is the primary food source. The way these muscles are built up is quite varied and their structure is very different from the muscles found on vertebrates. Those on cephalopods have a smoother structure and are made up of thinner muscle fibres than those of fish and terrestrial animals.

Which parts of a cephalopod are eaten depends quite a bit on the species and the size of the animal. In the case of octopuses, it is usually only the muscle mass from the arms, although occasionally the mantle is also incorporated in dishes. Virtually all parts of squid and cuttlefish—mantle, arms, tentacles, fins, and some of the innards—are considered edible. The very small species of cephalopods, as well as small specimens from larger species, are normally prepared whole, often without removing the innards. For example, in many countries around the Mediterranean, small squid and cuttlefish are grilled quickly at high heat and eaten in their entirety.

The innards also play a role in some cuisines. The liver from squid can be utilized as a fermentation aid to make products such as fish sauces, and the ink can be used to dye other foods black. The females of larger species of cuttlefish are also a source of nidamental glands, which secrete gelatinous substances that protect the eggs for reproduction. In Spain these are prepared as a special dish and are considered a great delicacy. Basically, the only parts of the animals that cannot be eaten in some form or other are all hard—the beak and the squid pen, which are made up of chitin, and the cuttlebone, which consists of a carbonate mineral.

There are three special challenges associated with promoting cephalopods as a food source. The first is that people whose culinary traditions do not include these animals might be reluctant to try them. The second is related to their strange appearance. With their many long arms and rows of suckers they look completely unlike the fish and terrestrial animals that many are used to thinking of as food. The third is a more practical worry about being able to prepare cephalopods so that they have an attractive and interesting mouthfeel and taste delicious. Overcoming the last challenge is a real problem because there must many who have been put off by eating badly prepared dishes such as deep-fried squid rings that were tough and rubbery.

We are going to take up these three challenges and try to turn those aspects of the cephalopods that at first glance are judged to be drawbacks into advantages. With regard to the first two challenges, it is our hope that the initial parts of the book have helped to demystify these creatures and shown them to be fascinating animals that deserve a place in many more food cultures. In order to meet the final challenge head on we will proceed to demonstrate how cephalopods can be prepared in a wide range of different ways and that they certainly do not need to have a tough mouthfeel, but can be tender, creamy, or crisp.

In the next chapter we will provide recipes that show how this can be done in practice. We will introduce a whole series of methods including drying, cooking, tenderizing, fermenting, smoking, sautéing in oil, deep-frying, grilling, breading, salting, and marinating. In principle it is possible to use cephalopods to make dishes that are as simple or as complex as one desires.

Our starting point is a discussion about them as raw ingredients—what to look for when buying them, how to clean freshly caught ones, and how to store them.

Finding a Source of Cephalopods

Apart from the challenges mentioned above, there is actually a fourth one for consumers who live in areas where cephalopods are not well-integrated into the culinary tradition—namely, finding places to buy them. Here is an overview of what to look for, whether they are fresh or frozen.

◨ Freshly caught squid at a street market in Naples.

In many countries it is virtually impossible to find live cephalopods in the same way as one can buy crabs or lobsters right from a tank. The cephalopods sold in fish stores, that appear to be freshly caught, are often frozen ones that have been thawed. But if one is lucky enough to come across really fresh cephalopods one is in for a treat. One can recognize the very fresh ones by their clean, slightly sweet smell, whereas if they are one or two days old they have already developed a strong, possibly unpleasant, fish-like smell. The colour of their gills should be bright and the flesh should have a firm, springy, slightly quivery texture. If the skin still has an iridescent sheen it is a yet another indicator that the taste experience will be positive.

Very fresh, even live, cephalopods can be eaten completely raw. Without a doubt, the best raw ingredients for a dish made with cephalopods come from freshly caught specimens that are cleaned only just before they are to be prepared. As is the case for fish, it is best to store fresh cephalopods on ice.

While the selection may be limited, it is generally easier to buy cephalopods in frozen form. Fortunately they freeze very well, even better than fish, because they contain relatively little fat and their muscle meat is fairly homogeneous. And unlike the meat from terrestrial animals, there is no blood in cephalopod meat, so there are none of the by-products that result from the break down of blood, which can leave an aftertaste. In some countries it is possible to buy dried or salted squid and cuttlefish, which need to be thoroughly soaked in water before use.

Cleaning and Storing Cephalopods

There is no need to be intimated by the prospect of cleaning cephalopods. Broadly speaking, all cephalopods are cleaned in the same way, although there are a few tricks to deal with the differences between them. How they are cut up depends to a certain extent on their size. The methods are described below. Since it may difficult to visualize and follow the written instructions, many readers will find it more instructive simply to find an appropriate video on the internet.

Squid and cuttlefish are cut up in the same way, taking into account the difference in how the pen and the cuttlebone are removed. Cuttlefish usually have a thicker mantle, thicker skin, and often a larger ink supply. If the whole body is to be stuffed or cut into rings, the squid should not be slit open to clean it. Sometimes it is easier to use a pair of kitchen scissors instead of a knife to cut squid and cuttlefish. The ink sacs can be removed carefully and kept for later use, including by freezing them. The following procedures apply to cephalopods that are already killed.

The Step-by-Step Method for a Cuttlefish

- Use your fingers to hold it near a fin and cut a small slit in the tough skin.
- Grasp the skin and pull it off, including that on the fins.
- Make a cut all along one side of the cuttlebone.
- Open it and find the innards located behind the cuttlebone.
- Being careful not to burst the ink sac, make a cut on the other side of the cuttlebone and pull away the mantle from the rest.
- Cut off or squeeze out the beak.

- Set aside the arms, tentacles, head, and innards, including the ink sac and nidamental glands (from females), for further use.
- Wash the mantle, the arms and tentacles thoroughly in cold water and rub the suckers, possibly with a bit of salt, to clean them completely.

The Step-by-Step Method for a Squid

- Grasp the arms and pull them away from the mantle, taking as much as possible of the innards with them. Look for the ink sac and set it aside if it is to be used later.
- Place the piece with the arms and the tentacles on a cutting board and cut off the arms.
- Cut off or squeeze out the beak.
- Wash the arms and tentacles thoroughly in cold water and rub the suckers, possibly with a bit of salt, to clean them completely. If necessary, cut off the biggest suckers that may have some hard chitin rings.
- Lay the mantle on a cutting board and with the back of a knife 'iron' it starting from the pointed end to squeeze out the remaining innards. Use your fingers to pull out the pen.
- Clean out the inside of the mantle with your fingers or a spoon.
- If the mantle is not to be stuffed or made into rings, it can be prepared more easily by cutting it open lengthwise, carefully pulling away the innards and tentacles, and then gently scraping it clean with a knife.
- Take hold of the skin at the top and pull it off. Sometimes a little salt on the fingers helps to grip it. This will also pull off the fins, which are used in some dishes.
- Set aside the innards, arms, and tentacles for later use.

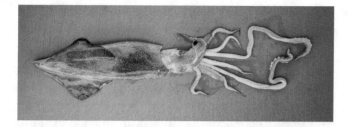

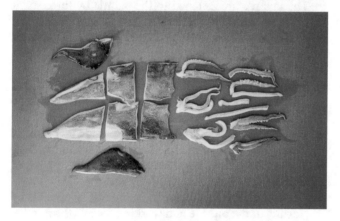

Large squid can also be prepared more easily by cutting through the mantle lengthwise along the pen in order to avoid damaging the innards. The pen can then be removed and the innards pulled out. On these large creatures it is quite easy to look through the innards to find the siphon and the muscles that are attached to it.

Regardless of how squid and cuttlefish are eventually going to be eaten, it can be an advantage to start tenderizing the mantle right away as part of the cutting up process. Using a sharp knife score the meat in a diamond pattern with the cuts about four to five millimetres apart. Pieces that have been tenderized in this way can be used immediately in sushi or *sashimi* or be stir-fried in a wok, in which case the scoring helps to ensure that the heat is evenly distributed over the whole surface.

The Step-by-Step Method for an Octopus

- Place the octopus on a cutting board and use a knife to make a cut above and under the eyes in order to remove and discard the eyes without cutting right through the flesh on the opposite side.
- If it is a larger octopus, turn the mantle inside out. If it is a small octopus, cut the mantle open lengthwise. This will make it easier to clean it. Try very hard not to puncture the ink sac, but if it happens wash off the meat.
- The innards are fastened to the mantle by three small pieces of flesh. Pull these away and discard them.
- Rinse the mantle thoroughly in cold water. The skin can be removed by grasping it at an edge and pulling it off, but this is strictly optional.
- Taking the piece with the arms, squeeze out or cut off the mouth part with the beak. Clean the arms thoroughly in cold water.
- Wash the suckers in cold water by rubbing them with a little salt to remove sand and any other unwanted bits.
- Dry all the pieces completely with a cloth or paper towels.
- Cut away the arms as close to the mantle as possible by starting at the outside and working toward the middle.

Storing

Cleaned fresh cephalopods should be stored in the refrigerator at the lowest possible temperature until they are to be used, ideally the same day. It is also possible to freeze them for later use, preferably in vacuum packs or in tightly sealed plastic bags. Alternatively, they can be frozen in a water bath, which will avoid freezer burn. They can keep for two to six months, depending on how well the temperature in the freezer is controlled.

It is best to thaw frozen cephalopods slowly in the refrigerator, but they can also be thawed more quickly by placing them in cold water. As is always the case with meat, some liquid will seep out when cephalopods are thawed because the ice crystals have destroyed their cell structure. It is necessary to take this loss into account when preparing them by cooking them more quickly so that the meat does not turn out mealy and dry.

It can be difficult to judge the quality of commercially frozen cephalopods, but as a general rule it is best to avoid any that have gone yellow or have signs of freezer burn.

Sautéed squid

Serves 4

Ca. 1 kg (2 1/4 lb) squid (*Loligo forbesii*)
1 dL (3/8 c) olive oil
1 or 2 gloves garlic
Oil for frying
1 organic lemon

1. Clean the squid and cut the arms and the tentacles from the mantle.
2. Cut the mantle into 3–4 cm (1 1/8–1 1/2 in) pieces and the arms and tentacles into pieces that are 5–6 cm (2–2 1/3 in) long. Dry well with paper towel.
3. Place the pieces in a dish with the olive oil. Put the garlic through a press and sprinkle on top. Set the squid pieces aside for at least 30 minutes before they are to be sautéed.
4. Sauté the squid pieces at high heat in a skillet, turning them over several times. When they have turned golden at the edges, remove them with a slotted spoon and place on paper towel to absorb the excess oil.
5. Season to taste with sea salt and freshly ground pepper.

To serve

Arrange the warm squid pieces and lemon wedges on a serving plate. Serve as a snack, perhaps with a little garlic mayonnaise for dipping.

🔲 Sautéed squid.

Tenderizing to Improve Texture

Because it can be tricky to prepare cephalopods so that they have a pleasant mouthfeel, one has to accept the risk that they might end up being tough and rubbery. Even the experts sometimes have this problem. This is where tenderizing comes into play, including when the meat is to be eaten raw. There are many ways to improve the texture of the muscle meat. In most cases, heat is used to tenderize, usually by cooking the meat in water, but also by deep frying and grilling it, or using a pressure cooker. Other methods include mechanical massaging, curing it with salt, acids, and enzymes, and freezing it. Sometimes two or more of these methods are used in combination.

But no matter how one goes about preparing them, one should be willing to accept that the structure of cephalopods is such that it will almost always require some effort to chew them. Of course, this is precisely one of the reasons for eating them! Another factor is the kind of dishes in which they are used. If they are to be very tender, it is necessary to cook and simmer some cephalopods for a long time, although one then risks ending up with less tasty, dry meat. If one wants meat that is succulent and more chewy, it should be cooked quickly. If one wishes to have something that is firmer, one might consider eating some kinds raw or almost raw, for example, in *sashimi* or octopus salad.

Octopuses invariably need to be tenderized by some means, otherwise they are not really edible. Some squid from the *Loligo* genus, for example *Loligo forbesii* and *Loligo vulgaris*, do not need to be tenderized, whereas the reverse is true for the jumbo flying squid (*Dosidicus gigas*).

Tenderizing with Heat

The preferred, or at least the most common, way to tenderize cephalopods, especially octopuses, involves heat and this is where the fun begins. It is much trickier to find the right balance between temperature and cooking time than it is for meat from terrestrial animals and fish. Cephalopod meat should be sliced into thinner pieces than other types of meat. Its taste is less pronounced and it more readily loses its juices and dries out.

As described earlier in the book, the complexity of cephalopod anatomy and their peculiar muscle structure have to be taken into account in order to end up with the right texture. The muscle fibres and connective tissue are strongly crosslinked in all directions and the meat has four times more collagen than that of finfish. Consequently, the outcome is very much affected by how, and for how long, the meat is heated. This is somewhat the same as the problem that one has when cooking other meats, namely, that the connective tissue and the proteins in the muscles each have their characteristic and different ways of reacting to heat. Salt and acid also have an influence on how quickly the proteins denature. For example, vinegar will stiffen the muscle proteins but at the same time promote the breakdown of the collagen, thereby releasing gelatine, which can dry out the meat. Salt seems to be able to counteract this effect but will affect the taste.

When muscle meat is heated the muscle fibres contract, the proteins partially denature, and the meat becomes firmer rather than more tender because the connective tissue has not had sufficient time to break down completely into gelatine. Hence, the guiding principle when treating all types of cephalopods with heat is that they should be cooked either exceptionally quickly or else simmered for hours. Anything in between these extremes, especially in the case of octopuses, risks making them tough instead of tender.

Heating the cephalopod meat really quickly leaves it tender and juicy because the proteins have not had time to denature fully and the collagen has been softened somewhat. If the cooking time is a little too long, the meat becomes tough because the collagen contracts in all directions and the meat loses its liquid.

The bonds in cephalopod collagen break down only after they have been cooked for a very long time. The resulting small pieces of insoluble collagen are then converted irreversibly into water soluble gelatine, making the meat more tender and possibly juicy. It is important to keep in mind that this is a hydrolytic process and hydrolysis takes time, especially at lower temperatures.

Scientific measurements of the effects of heating cephalopod meat have shown that the collagen slowly starts to turn into gelatine and dissolve in the water at about 40° Celsius, and somewhat more quickly in the arms than in the mantle. But it is only when the temperature goes above 60° Celsius that the process really picks up speed. This is why most recipes suggest that cephalopods should be cooked or at least simmered at around the boiling point of water. Nevertheless, it is likely that a somewhat lower temperature might be better.

Connective tissue can contain a fair amount of fat, which can help to make meat more tender, although the effect is slight. Fat is much more important as a source of taste, but there is too little of it in cephalopod meat to make a difference to its taste or texture.

High-speed cooking can be done by plunging the cephalopod in either boiling water or very hot oil for no more than ten to twenty seconds or by placing the cephalopod in a sieve and pouring the hot oil over it. This leaves the meat tender and succulent, while preserving a little of its bite. For example, preparing the mantle of a cuttlefish in this careful way leaves it almost raw on the inside.

◧ Cooking a whole *Octopus vulgaris* in water.

There is no doubt that the preparation of octopuses poses the most challenges and there are just as many opinions about how to do this as there are cooks and cookbook authors. Some Spanish chefs maintain that octopuses can be cooked only in copper pots; the Italians say that one should put two corks in the cooking water; the Greeks recommend beating the octopuses against a rockface; and Japanese cooks suggest massaging the octopuses by hand and rubbing them with salt and grated *daikon* or pounding them with a small wooden club.

Octopus vulgaris is native to the Mediterranean and many parts of the Atlantic and is the most commonly eaten species. It can be as much as two metres long and weigh up to twenty kilograms, but those that are usually caught are much smaller, weighing a couple of kilograms. All parts of the arms can be eaten if they are tenderized, as can the mantle, although it can be somewhat chewier. The arms are almost pure muscle and have a lot of very clean, white meat on the inside. It is surrounded first by a slightly tougher layer of collagen and then by a thin layer of skin on the outside. Some cooks rub off these layers before cooking an octopus but others prefer to leave them on, which results in a sticky, gelatinous surface. The layer of skin dries out almost completely if the octopus arms are grilled or cooked without water.

Tenderizing Using Pressure and Sous Vide Techniques

Food preparation is normally carried out at standard atmospheric pressure by making changes to the temperature, while leaving pressure constant. But as most physical and chemical processes are dependent on both pressure and temperature, it is possible to shorten cooking times by placing the ingredients in a pressure cooker. For example, water under pressure boils at a temperature higher than 100° Celsius. This makes it possible to heat the food to a temperature beyond the normal boiling point, even if water is present.

We have successfully cooked large octopuses in a pressure cooker at an atmospheric pressure of 1.7, at which the water boils at 117° Celsius. The cooking time was shortened from 18 minutes in an open pot to 10 minutes in a pressure cooker. The octopus was then allowed to steep in the cooking water for a further 18 minutes in the pot or for a similar length of time in the pressure cooker until the pressure had returned to normal. Cooking reduced the weight of the animal. For example, a specimen that initially weighed 1.6 kilograms was cooked in 3 litres of water with 30 grams of salt and lost half its weight.

While sous vide preparation of squid and cuttlefish leads to a demonstrable effect of their tenderness, it serves to reinforce the idea that a short cooking time results in the best overall sensory experience. When cooked for a very long time, the surface of their meat ends up overdone and the texture is too soft and uniform. The characteristic 'bite' of the cephalopod meat is lost and it becomes mealy and bitter with a somewhat 'fishy' smell.

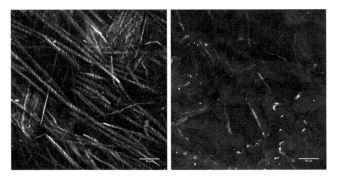

◘ Micrographs of the collagen structure in the mantle of *Loligo forbesii* in its raw state (on the left) and after sous vide preparation (on the right). Cooking has opened up the tightly interwoven collagen structure, which tenderizes the muscle meat. The scale at the bottom right is equivalent to 50 μm.

Mechanical Tenderizing

In David Gelb's film *Jiro Dreams of Sushi*, there is a scene in which the old sushi chef Jiro Ono's oldest son Yoshikazu massages on octopus with his hands for more than half an hour before it is ready to be cut up to make first class sushi. What this does is loosen the muscle fibres in relation to each other. Rubbing octopuses by hand is probably the most elegant and least damaging way to tenderize them, but without a doubt it is also very time consuming and monotonous.

A more modern technique involves the use of a tumbler in which the arms are whirled around for hours in a little salt water. When the process is finished, the arms have curled up somewhat, but they have become considerably more tender, even to the point where they can be eaten raw.

In some countries in the Mediterranean, one can see the fishers holding the octopuses by an arm and bashing them against rocks and cliff faces, an activity that has a dual purpose. First of all, hitting tenderizes them by separating the muscle fibres from each other, which is exactly what happens when meat is pounded. Secondly, because of their distributed nervous system, it is not always easy to kill these animals, not even by cutting away the body just under the eyes, but the impact has enough force to do the job.

An octopus must be beaten seven times.
An old Greek proverb

A less brutal way to tenderize the meat is to score it with a knife in a diamond pattern to cut through some of the muscle fibres. This method is used on squid meat that is to be fried very rapidly in oil or doused with very hot oil. One can also observe sushi chefs score the underside of the mantle of raw squid and cuttlefish when they are to be made into *sashimi* or sushi (*ika sushi*). Similarly, small cuts can be made in the edge of cut-up pieces of cooked octopus arms for sushi (*tako sushi*).

The ultimate way to tenderize cephalopods mechanically is to grind them up extremely finely, virtually pulverizing them, and then adding binding substances and pressing them together into blocks or tubes. This is done, for example, in Japan to produce *surimi* from the jumbo flying squid (*Dosidicus gigas*). The taste of the *surimi* is enhanced with seasonings such as soy sauce, other sources of umami, spices, and sugar.

Freezing

Freezing causes some of the water in the muscle tissue to crystallize, which helps to burst the cells and tenderize the meat. The drawback, of course, is that the meat loses some of its liquid when it is thawed, so that it ends up drier and less tender. Our experiments have shown that the texture of squid and cuttlefish does not change noticeably after it has been frozen, whereas it really affects octopuses negatively. Sadly, as most consumers would rarely be able to purchase fresh octopuses at a fish market, they will in almost always have to resort to ones that have been frozen for shorter or longer periods of time.

Tenderizing with Enzymes

The muscle meat from cephalopods, like that from terrestrial animals and fish, can be tenderized by the action of specific enzymes. It is actually possible to do this using the enzymes found in the squid's own ink, in which case one should use fresh ink that has not been heated. Enzymes from fruits such as papayas, tomatoes, and pineapples are also able to tenderize the protein fibres in cephalopod meat. Using a microscope one can see that the enzyme bromelian from pineapple juice opens up the muscle structure of squid, making it softer.

◘ Tenderizing squid meat from *Loligo forbesii* with miso and with pineapple juice.

Tenderizing Using Acids and Salt

There is no agreement about how different acids affect the tenderness of cephalopod meat when it is slow cooked. As a result of his own experiments, the renowned food authority Harold McGee has come to the conclusion that one should forget all about tenderizing using acids, such as acetic acid from vinegar, citric acid, and lactic acid. Instead one should focus on using salt, even though it may not have a positive effect on the taste of the meat once it is prepared.

McGee uses acetic acid as an example. When heated, it causes more of the collagen to break down and convert to gelatine. He writes that this has the opposite effect of what is desired, because the meat ends up too dry and fibrous. So even though the acidic environment increased the extent to which the stiff connective tissue was broken down, it also degraded the resulting gelatine, leaving the muscle fibres without any lubrication and even more fibrous.

According to McGee, salt can have the opposite effect by tenderizing the muscle fibres. He describes the process in his well-known New York Times column, *The Curious Cook*. He brined some octopus arms in a 5 percent salt solution and then simmered them in water at 90° Celsius for four to five hours. They ended up less fibrous and not noticeably salty, but the effort did not make the most of their flavour. His solution was to blanch the octopus arms for about thirty seconds in boiling water and then allow them to simmer in their own juices in a covered dry pan at 90–95° Celsius for four to five hours.

Our own experiments have shown that we achieved the best results from placing an octopus in a 7 percent brine (70 grams of salt in 1 litre of water) for 5 minutes, cooking it for 18 minutes in its own water, and then letting it rest in the cooking water for a further 18 minutes.

Cephalopods Can Be Both Fermented and Used to Ferment

The innards of squid and cuttlefish, in particular the liver, contain some very aggressive enzymes (proteases) that are able to break down proteins. In southeast Asia these innards are used industrially to ferment fish, molluscs, and crustaceans and to produce fermented sauces. One example is *ishiru*, a Japanese specialty sauce made with fermented sardines, other small fish, and cephalopod innards, especially those from the Japanese flying squid, *Todarodes pacificus*.

Shiokara is a Japanese term that can be loosely translated to mean something that is salted and spiced. It designates products that are fermented for up to months at a time with the help of the liver from cephalopods and significant amounts of salt (10–30 percent of the total mass). During the fermentation process, the enzymatic action leads to the formation of a long series of breakdown products such as peptides and free amino acids, which among other things impart strong *kokumi* and umami tastes. The high salt content prevents the growth of undesirable microorganisms in the course of the long fermentation period. In Korea *shiokara* is called *chokkara* and is usually made from fermented fish, shrimp, and molluscs. It is often served with *kimchi*.

Fermented Cephalopods—Japanese Style

Several species of squid and cuttlefish are used in Japan to prepare *shiokara*. In the traditional recipe for *ika no shiokara* the mantle and the fins are fermented in the animal's own innards. Fermentation is normally carried out during the colder months of the year and can take up to half a year, with the end taste becoming stronger over time.

The enzymes from the *shiokara* break down the proteins in the cephalopod meat, thereby tenderizing it. This effect can be modified by adding a paste made from malted rice, which contains certain other enzymes (oryza cystein protease) that suppress the action of the *shiokara* proteases and also add a touch of sweetness. As a result, the fermented cephalopods are a little firm, have a pleasant mouthfeel, and their texture is somewhat like that of raw meat.

Ika no shiokara is most often made from Japanese flying squid (*Todarodes pacificus surume* in Japanese). A more refined, specialty version, *hotaru ika no shiokara*, uses small firefly squid (*Watasenia scintillans*).

The taste of *ika no shiokara* is primarily derived from the liver in *shiokara* and the breakdown products from the fermentation process, especially free amino acids such as glutamic acid (glutamate), which contributes umami. A longer fermentation period and subsequent curing time result in an increasingly strong taste of umami. It is thought that the liver may also contain microorganisms that release taste substances.

Ika no shiokara can be seasoned with *shichimi* (a blended Japanese spice that contains chili), grated *yuzu* rind, sweet rice wine, or *wasabi*.

Ika no shiokara has a slightly sticky, soapy mouthfeel, a lasting aftertaste, and is, of course, very salty. It is eaten as is in small portions or on top of a bowl of rice. It can also be served as a small side-dish with a meal, or as an appetizer in a bar where it is washed down with a glass of whiskey or sake.

Preparing *ika no shiokara*

To prepare *ika no shiokara* from scratch it is necessary to obtain very fresh, preferably live, small squid with their innards intact.

ℹ *Ika no shiokara*

Fresh squid

Salt

1. Clean the squid.
2. Carefully remove the liver sacs from the innards without damaging them.
3. Place the liver sacs in a lot of salt and leave them for a couple of hours.
4. Clean off as much of the salt as possible.
5. Remove the thin membranes around the liver sacs or cut a hole in the sacs and scrape out the insides, which are then crushed or chopped finely and mixed with salt. This now constitutes the marinade, *shiokara*, that can be used to ferment the cephalopod meat.
6. Remove the fins from the mantle.
7. Carefully tear off the outer coloured skin by grasping the skin at the top of the mantle. If necessary, trim off any remaining bits.
8. Slice the mantles open along the side and lay them flat on a cutting board.
9. Cut the mantles into strips that are 5–10 mm (1/5–2/5 in) wide.
10. Place the strips in the marinade, put them in an airtight container, and leave them in the refrigerator for two days.
11. When serving, season according to taste with salt and, possibly, some *miso*, sake, or *mirin*.
12. Eat within two days.
13. Normally the arms, tentacles, and fins are not used to make *ika no shiokara*, but can be used to make other dishes.

◘ *Ika no shiokara*: squid (*Loligo forbesii*) tenderized in the enzymes from its own innards and with a little ink.

Dishes with Ink

Cephalopod ink is used in food preparation to add dark and black tones and to help to tenderize the meat. It is found in the ink sacs that can be separated out carefully from the rest of the innards, with *Sepia* being the best source. It is also possible to buy commercially prepared ink in jars or pouches and in powder form, but it is important to remember that it has been pasteurized and no longer contains active enzymes.

Cephalopod ink has been used for centuries in many traditional cuisines, especially in Mediterranean countries and in Asia. It is now coming into its own elsewhere as a source not just of colour, but also of interesting, slightly salty, and fish-like taste nuances, and has been referred to as 'the new black.' Apart from being used as in ingredient in virtually any type or shape of pasta, it is also common in rice dishes. When added to meat dishes, sauces, and baked goods such as cakes, bread, and *grissini*, it can create a startling visual appeal.

A classic dish made with cephalopod ink is *calamares en su tinta* or squid cooked in their own ink.

ℹ️ *Calamares en su tinta*

Serves 4

1 kg (2 1/5 lb) squid or cuttlefish, for example, small *Sepia*

3 large onions

2 cloves garlic

2 dL (3/4 c) dry white wine

6 Tbsp extra virgin olive oil

1 ink sac

Salt and pepper

1. Clean the cephalopods, removing the innards, mouth, and eyes and rinse them thoroughly in cold water. Be sure to preserve the ink sacs.
2. Cut the mantle into rings and the arms and tentacles into smaller pieces.
3. Place the ink sac in a glass with a few drops of water and crush it with a spoon.
4. Peel the onions and cut them up into odd sized pieces.
5. Peel the garlic cloves and chop them finely.
6. Pour the olive oil into a skillet and heat it.
7. Sauté the onions and the garlic in the skillet for about 10 minutes.
8. Add the cephalopod pieces and sauté for a further 10 minutes.
9. Add the wine and let it come to a boil for a couple of minutes.
10. Add the inky water and, if desired, thicken the sauce to an appropriate consistency.
11. Season with salt and pepper.
12. Can be served with cooked rice.

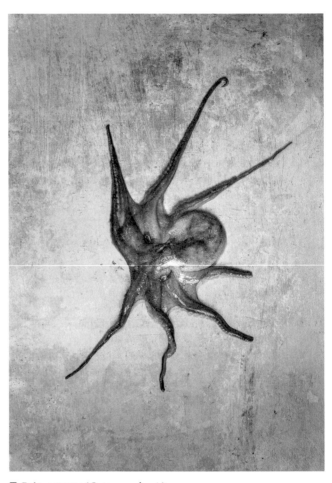

❏ Baby octopus (*Octopus vulgaris*).

Cephalopod Cuisine—Its Global Reach

In this chapter we are going to explore the various ways in which cephalopods have been incorporated into the cuisines of virtually all parts of the world, going right back to ancient times. We will provide many examples of how they can be prepared to make traditional dishes as well as inventive new ones. The culinary techniques needed to do so are based on an understanding of cephalopod anatomy and of which parts of the animals are best suited for eating, as described earlier in the book. It will become apparent that although the particular properties of cephalopods pose challenges, they can also be used to advantage.

The recipes we are presenting have been selected according to three criteria. One is to give an impression of how cephalopods are used globally in regional and national food cultures. Another is to demonstrate that cephalopods are a very versatile protein source that should be seen as an important alternative to fish and terrestrial animals. And a third is to emphasize that it is actually quite easy to prepare them in a whole range of ways to make many different dishes.

While most of these recipes have been developed independently and tested in conjunction with the writing of this book, some have been inspired by cookbooks that are listed in the sources at the back of the book. We have also gleaned ideas from the wealth of information that is found on the internet, adding a truly international flavour.

Many of the recipes are very easy to follow. But we decided also to include a few that demand more culinary expertise and are more labour-intensive in order to demonstrate that cephalopods are also able to meet the challenges posed by fine gastronomy.

From Ancient Times to Now

Cephalopods have found a place in European food cultures for millennia. The ancient Greeks and Romans caught octopuses in pots and all types of cephalopods were served at large banquets. Octopuses were given as gifts on the name day of a newborn child and they were also regarded as an aphrodisiac.

The oldest cookbook that has come down to us from antiquity is *De re coquinaria* (*On the Subject of Cooking*), which is attributed to the Roman gourmet and *bon vivant* Marcus Gavius Apicius (25 BCE–37 CE), although it is doubtful that he was the actual author. It includes a recipe for octopus seasoned with pepper, lovage, ginger, and

"Octopus strengthens the sex drive, but it is tough and indigestible."

Alexis (Greek poet, ca. 375– ca. 275 BCE)

garum (a fermented fish sauce). Recipes from the Middle Ages are few and far between. One can be found in a manuscript dating from about the middle of the 1300s written by an anonymous scribe living the Kingdom of Aragon. This Catalan recipe is for an octopus stuffed with its own arms together with spices, parsley, garlic, raisins, and onions and then grilled over charcoals or baked in an oven.

After the invention of the printing press in the time of the Renaissance, cookbooks started to be more readily available and they frequently contained recipes for cephalopods. Examples are a Catalan recipe for baked octopus and Italian ones for cooked, fried, and marinated octopus. In Galicia in the northwest corner of Spain, octopuses have been eaten since earliest times and dried cephalopods were brought inland by coastal traders. To this day, octopus is an important ingredient in the Galician kitchen, where the signature dish of the region is *pulpo a la Gallega*, for which there is a recipe later in the chapter.

At this time, cephalopod cuisine probably features most prominently in the Far East, especially in China and, even more so, in Japan. Cephalopods are also incorporated into traditional Chinese medicine. Japan might well be the country in which they, octopuses in particular, are most highly prized. The Japanese eat more octopus per capita than the people of any other nation and their needs can be fulfilled only by importing them in massive quantities.

The octopuses found around the island of Awaji, located in the Seto Inland Sea between Honshu and Shikoku in eastern Japan, are reputed to be the very best. It is thought that those caught in the summer months owe their superb taste to their being able to feast on an abundance of crabs and shrimp. A special local food culture has grown up around the octopus (*tako*) and it features in a wealth of recipes. The most famous is probably *tako-yaki*, a type of dumpling made by wrapping dough seasoned with ginger and spring onions around minced, cooked octopus meat and adding *tempura* flakes. The dough is then fried in special molded metal pans and served with a variety of dipping sauces, one of which is *dashi*. *Tako-yaki* is available everywhere in the area around Osaka, where it is a popular street food. Octopus is actually so well-loved in this part of Japan that it is celebrated on a special *Tako* Day on the 2nd of June.

Cephalopods, especially squid and cuttlefish, are also eaten completely raw or grilled just a little and made into *sashimi*, which is dipped in soy sauce, *wasabi*, or *ponzu*. Lightly cooked octopus arms are also served as *tako* salad. At the other end of the spectrum one finds dishes with slow-cooked octopus, such as *nimono*, which has been simmered for hours in *dashi*. These are often served with vegetables on the side.

Raw or Almost Raw

It is to be expected that the preparation of a particular ingredient varies greatly from one country to another and is reflected in their recipes and food cultures. These differences can be seen clearly by comparing the approaches in the three countries that are the biggest consumers of cephalopods: Spain, Italy, and Japan. But it may come as a surprise that, whereas in Japan they are eaten raw or prepared in a very minimalist way, in Spain there is no similar tradition of food from the sea in raw form. The approach in Italian cuisine falls somewhere in between these two opposites. Like the Spaniards, Italians deep-fry and grill cephalopods, but there are also numerous, often regional, dishes prepared with raw fish and shellfish (*pesce crudo*), as well as cephalopods.

Many types of squid and cuttlefish can be eaten raw, or almost raw, provided that one is sure that they are really fresh or have been frozen shortly after they were caught. Nevertheless, some cephalopods can be infected with parasites and if in doubt it is best not to eat them completely raw. The standard rule of thumb for raw fish—namely, to freeze them to −20 °C for at least 24 hours, and preferably 72 hours—applies for cephalopods as well. The special Korean dish *san-nakji*, described in an earlier chapter, is made with the arms of a still living octopus, but this must be regarded as taking the concept of raw to extreme lengths.

Japanese cuisine is replete with dishes made with all kinds of marine products, with sushi and *sashimi* probably being the most widely recognized example. *Sashimi* is often prepared from the same pieces of cephalopod meat as sushi, although there are some variations in the selection.

To go from a completely raw dish to one that is almost raw requires very little effort. It is usually a simple matter of marinating cephalopod pieces for a short period of time using salt and acids such as rice vinegar and citrus juice. South American *ceviche*, in its many varieties, and Italian *pesce crudo* are classic examples of such dishes.

takotsubo ya
hakanaki yume o
natsu no tsuki
octopus traps
fleeting dreams
under the summer moon
Matsuo Basho
(1644−1694), haiku
written in the summer
of 1688 when spending
the night at Akashi on
the Seto Inland Sea

Cephalopod Sushi

Octopuses, squid, and cuttlefish are used to make classical *nigiri-sushi*, which consists of a suitable piece of cephalopod meat placed on top of a small, hand-shaped ball of sushi rice. *Tako-sushi* is normally made using the arms of *Octopus vulgaris* that have been cooked to tenderize them. *Ika-sushi* is made using the mantle of both squid and cuttlefish. In most cases this meat is completely raw, although it has usually been frozen before use.

When an octopus is cooked its surface colour changes to a reddish-violet, while the meat inside the arms becomes chalk white. The pieces are cut to showcase these colour contrasts, which are an important esthetic aspect of how *tako-sushi* is presented. The best sushi is made from the thickest arms.

◘ *Nigiri-sushi* made with cuttlefish, *ika-sushi*, and with octopus, *tako-sushi*. The octopus meat is fastened to the rice ball with a thin strip of seaweed (*nori*).

The thickness and firmness of the muscle meat of different species of squid and cuttlefish varies a great deal, ranging from thin and tender to that which is thick and tougher. Squid, such as *Loligo*, can be eaten raw, while cuttlefish (*Sepia*) are often cooked very quickly before use. The cephalopods with the thickest mantle are the most suitable as their meat tends to resist a little when bitten into and it is pleasantly sweet and creamy, all of which provides a subtle contrast to the slightly sweet and sour sushi rice.

Chirashi-sushi (literally meaning 'scattered sushi') is a colourful dish in which fish, shellfish, and vegetables are scattered on top of a bowl of sushi rice and often sprinkled with finely chopped *nori*, flying fish roe, or toasted sesame seeds. It can be made with cephalopod meat that has been prepared for *nigiri-sushi*. For a particularly elegant presentation, pieces of squid are sliced into long thin strips that resemble noodles (*ika-sōmen*).

Squid and Cuttlefish *Sashimi (Ika-Sashimi)*

Squid, for example various species of *Loligo* and Pacific flying squid (*Todarodes pacificus*), can very easily be eaten raw as *sashimi*. They are prepared in exactly the same way as for sushi and one can introduce a little variety by scoring the pieces to form a number of different patterns. The *sashimi* should be dipped in Japanese soy sauce that has a bit of *wasabi* dissolved in it. Squid can also be made into an appetizer by combining them with avocado and fish roe.

 Ika-sashimi with avocado and lumpfish roe

Serves 4
400–500 g (about 1 lb) cleaned squid mantle
1 avocado
Wasabi
2 Tbsp cooking sake
100 g (3 1/2 oz) lumpfish or flying fish roe
Lemon juice

1. Slice the mantles open along the side and lay them flat on a cutting board with the exterior side up. Without slicing all the way through, carefully score the meat in a striped or diamond pattern.
2. Cut the mantle into pieces that are of a size suitable for *sashimi*.
3. Peel and pit the avocado, cut it up into thin slices, and arrange them on four serving plates.
4. Mix together the *wasabi* and the sake and brush the avocado with the mixture.
5. Toss the squid pieces with the fish roe, ensuring that there is some on each of them, and place them on top of the avocado.
6. Just before serving, sprinkle with a few drops of lemon juice.

This very simple *sashimi* appetizer can be made even more elegant by topping the arrangements with a few very small cubes of ripe mango.

A Cephalopod Feast in London

One of the best Japanese restaurants in England is located in a little side street in London's Mayfair district, just a stone's throw from the fashionable boutiques on New Bond Street. The only outward sign of its presence is a subdued dark red banner inscribed with the symbol 生 (Umu). One of us (Ole), together with an acquaintance, Sakiko Nashihare, were to meet Umu's head chef, Yoshinori Ishii, one of London's most feted Japanese chefs, who had prepared a lunch for us. I was about to sample the most perfect squid *sashimi* that I had ever tasted. It is not without reason that Umu has been awarded two Michelin stars.

As soon as we stepped inside we realized that we were in for a multi-sensory experience—the decor is understated and elegant, featuring dark wood, a few simple decorations, and soft, pleasant lighting. Ishii-san has undergone classical training not just in the culinary arts, but has also achieved mastery of such crafts as *ikebana*, calligraphy, and Japanese ceramic techniques.

When we were exchanging greetings, I asked Ishii-san about the meaning of Umu. His answer indicated that the name was chosen with great care and that *umu* is a Japanese word for 'nature.' Like many Japanese expressions, however, it has a variety of meanings, which refer to different aspects of an object or a concept. For example, just as *hashi* translates both as chopsticks and as a bridge, *umu* can be interpreted simply as 'nature' and also to indicate a process—'the path to nature or to that which is natural.'

We were to eat our meal at the counter in front of the kitchen so that it would be possible for us to see how Ishii-san prepared the food, while also giving us a unique opportunity to carry out a conversation. Ishii-san said that when he first arrived in England in 2010 is was very difficult for him to obtain fish that were freshly caught, preferably the same day. But, little by little, he built up a network of fishers who bring him supplies from Cornwall and Scotland. Quite surprisingly, he told us that he had come to prefer several of these really fresh local products to fish that are delivered from Japan. We started the lunch with *sashimi* made from fresh brill from Cornwall. Normally, when brill or turbot are used to make sushi, the fish is aged on ice for two days. But Ishii-san explained that if the brill is to be cut into paper-thin slices for *sashimi*, it must be completely fresh as it otherwise has a texture that is too soft.

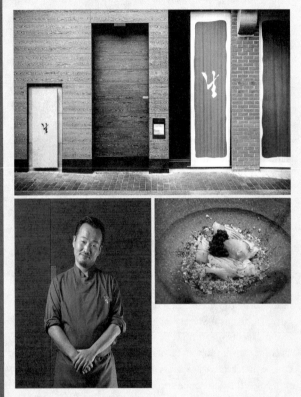

◘ Restaurant Umu in London, head chef Yoshinori Ishii, and one of his cuttlefish dishes.

Next we were served *ika sashimi* made from the mantle of a cuttlefish. It was obviously a large specimen as the piece we ate was almost one centimetre thick. It was lightly scored in a crosswise pattern so that the glistening, smooth morsel twisted slightly. This gave a beautiful esthetic effect that was more appealing than if it had lain flat on the plate. Ishii-san freezes the *ika* before he cuts it up for *sashimi*. He feels that this makes the meat sweeter and creamier, an idea that led to an interesting discussion. This was borne out in practice, as the *sashimi* was truly very creamy and had a perfect, soft texture. Amazingly, after chewing only a few times, there was virtually no tough tissue left for the teeth to work on, which is not often the case when squid is eaten raw.

We spoke a bit about the taste, but Ishii-san, a very humble, modest individual, articulated his worries about not being able to express himself adequately in English. Like the true craftsman that he is, he simply pointed to his chest and said "I have only my hands and my sense of taste." I interpreted this as an expression that encompassed a wealth of experience and a deep understanding of the quality of raw ingredients, of workmanship, and of the culinary arts.

Cefalopodi Crudo

A popular Italian dish, *pesce crudo*, could actually more correctly be called *cefalopodi crudo*, as it is usually made with squid and cuttlefish mantles. How these are sliced has a major influence on their mouthfeel. For example, the thin squid mantle is often cut into long strips, while the thicker cuttlefish one may be cut on the diagonal. These raw pieces can be prepared using different marinades.

The raw cephalopod meat presents a wealth of culinary possibilities. Combining it with shellfish such as crab or lobster results in a fine, complex taste experience. An interesting texture effect is created by putting together finely sliced tender, creamy cephalopod meat with crisp raw sunchokes that have been dyed black using cuttlefish ink. Another option is to mince the mantle of large cuttlefish to make a tartare-like dish.

Cephalopod 'fettucine' with lobster tails, roe, and dried lime

Serves 4

200 g (7 oz) piece of squid or cuttlefish mantle

8–10 very ripe cherry tomatoes

Brine made from 1 L (32 fl oz) of water and 50 g (3 1/3 Tbsp) sea salt

2 fresh lobster tails

Zest and juice from 2 limes

4 Tbsp *ponzu*

4 Tbsp olive oil, for the marinade plus more for sautéing

50 g (1 3/4 oz) salmon or other fish roe

50 g (1 3/4 oz) finely cut up pieces of peeled lemon

1 dried lime (possibly black lime, *loomi*)

1. Freeze the cephalopod meat for about a hour to make it easier to slice.
2. Blanch and peel the cherry tomatoes and set aside.
3. To make the brine, warm the water and salt mixture until the salt has dissolved and then allow it to cool to 5 °C (41 °F).
4. Cut the lobster tails in half, remove the shells and clean out the intestines.
5. Place a skewer lengthwise through the four lobster tail pieces and immerse them in the brine for 5 minutes.
6. Place the piece of mantle on a cutting board with the interior facing up. Slice it into long thin, even strips resembling fettucine. Set aside in a bowl.
7. Sieve the tomatoes to remove the seeds. Mix the resulting tomato pulp with the lime zest and juice, *ponzu*, and olive oil.
8. Pour the marinade over the cephalopod 'fettucine' and allow it to stand for at least 15 minutes in a cool place.
9. Heat a little olive oil in a skillet and quickly sauté the lobster tails.
10. Holding one end of the skewer, wrap the 'fettucine' around each lobster tail. Carefully remove the skewer.
11. Arrange each lobster tail on a plate, garnish with the salmon roe and lemon pieces, and drizzle with the remaining marinade.
12. Grate a little of the rind of the dried lime over them and serve.

◘ Lobster tails wrapped with 'fettucine' from long-finned squid (*Loligo forbesii*).

ⓘ Cuttlefish *tartare* with avocado, pistachios, and lime

Serves 4

150–200 g (about 1/3 to 1/2 lb) mantle meat from a
    large cuttlefish (*Sepia officinalis*)

4 tsp avocado oil

2 ripe avocados

Salt to taste

Zest and juice from 1 organic lime

10 g (1/3 oz) pistachio kernels

1. Freeze the cuttlefish meat for about a hour to make it easier to slice.
2. Slice the mantle meat into long, thin strips and then mince them as finely as possible.

3. Place the minced squid in a bowl and stir it until it starts to stick together slightly. Add a few drops of avocado oil. In order to preserve the clean, delicate, and unique taste of the cuttlefish, forego salt.

4. Cut the avocados in half, pit, and remove the peels. Slice the halves into thin pieces and arrange them in rows on a chopping board.

5. Neaten the slices by trimming off the tips, sprinkle them with a little salt and drizzle the lime juice over them.

6. Lightly toast the pistachios in a skillet and crush them in a mortar.

7. Using a spatula, transfer the rows of avocado slices to four plates. Divide the squid *tartare* into four portions and place a small egg-shaped mound on the avocado slices. Sprinkle the crushed pistachios and lime zest on top.

8. Decorate each plate with a few drops of avocado oil and serve.

◘ Squid *tartare* with avocado, pistachios, and lime.

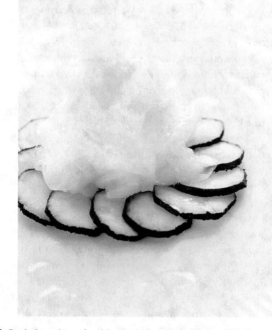 Cephalopod *royale* with sunchokes in squid ink

Serves 4
8–10 long, somewhat thin sunchokes
2 Tbsp squid ink
300 g (2/3 lb) piece of cuttlefish or squid mantle meat
Zest and juice from 1 organic lemon

1. Rinse and peel the sunchokes, pat them dry with paper towelling. Put the squid ink in a plastic bag and add the sunchokes, turning to coat them. Remove the sunchokes and place them on parchment paper to dry, moving them around a few times to allow them to dry evenly. Alternatively, dry them in a dehydrator or in the oven for about 30 minutes.
2. Just before serving, slice the sunchokes with a mandoline or a knife and arrange the slices on a plate.
3. Place the piece of mantle on a cutting board with the interior facing up. Slice off flakes without slicing all the way through the exterior skin side, thereby avoiding the tougher part of the mantle.
4. Using your fingers, shape the mantle flakes into a mound and place it on the sunchokes. Grate a little lemon zest on top, drizzle with a few drops of lemon juice, and serve at once.

Cephalopod *royale* with sunchokes dipped in squid ink.

In Search of Cephalopods in an Italian Street Market

Porta Nolana, the old street market in Naples, gets its name from a medieval city gate that dates from the fifteenth century. It was built to allow access to Naples from the road connecting it to the nearby city, Nola. The remnants of the old gate are now worn and blackened from air pollution, but the two large bastions are reminders of its once important function as a defence against unwelcome intruders.

While there is also a local train station at Porta Nolana, it is especially famous as the site of a colourful street market. The market stretches out along a few streets just behind the large Piazza Garibaldi. Here one finds everything from food to clothes, shoes, and toys.

One of us (Ole) had stumbled across this market many years ago when overnighting in Naples. To this day, I remember it as one of the worst experiences of my life. It was in August and the heat was just about unbearable for a northerner. The racket from outside with cars honking their horns all night was so penetrating that I did not sleep a wink. But I clearly recall wandering around in the streets in the early morning after a sleepless night and suddenly finding myself at the Porta Nolana market. Fish and shopkeepers were literally plonked down on the dark, dirty cobblestones. Despite its overall grubbiness, the effect of the atmosphere, the noise, the smells, and the foreign appearance of the brightly dressed Neapolitans was unbelievably invigorating. Now, many years later, I had come back to see whether I could come to appreciate more fully this fantastic market and its people, not least of all, in the hope of seeing some cephalopods for sale.

I was not disappointed. The smells and sounds had not changed and they helped to evoke vivid memories of my earlier visit. Nor was I disappointed with the selection of various types of cephalopods that had all been caught during the night in the Bay of Naples and brought to the market early in the morning. There were live specimens of the white spotted octopus (*Callistoctopus macropus*), which is commonly found in the shallow waters of the Mediterranean. This species is characterized by having very long arms connected by a shallow web. The first two arms are longer than the rest and can be up to about a metre in length. Other live specimens on display were large common octopuses

(*Octopus vulgaris*) and small curled octopuses (*Eledona cirrhosa*). A customer and a vendor were busy picking through the helpless *Eledona* lying in a tub of water. I supposed that they were trying to find the liveliest ones. Another vendor held up an *Octopus vulgaris* to show me that the creature was actually still alive. It wriggled its arms and used its suction cups to check out his hand. Generally, most of the cephalopods for sale were dead, although they looked to be very fresh.

■ Porta Nolana market in Naples.

There was also a large assortment of common cuttle-fish (*Sepia officinalis*). A few of them were still alive and moved around lazily in a little bit of water in a tub. Others were dead and arranged neatly with their arms folded in under the main body where their characteristic zebra-like stripes were clearly visible on their backs. One could also buy other species of cuttlefish, both large and small, some with splotches of ink oozing out over their pale bodies. Of course, there were also squid, but again no live ones, partly because it is almost impossible to keep them in water that is not flowing. Some of them had a beauti-ful dark purple appearance and others had grey skin with pretty dark red spots.

> Sadly, I could not buy any of the cephalopods, so I had to content myself with taking them in with my eyes and looking forward to eating some for dinner at a restaurant down by the shore. And I was not disappointed. I had both a warm octopus salad and whole, steamed curled octopuses together with squid mantles in a fantastic Neapolitan seafood soup.

Cephalopods in Marinades and Sauces

Marinades and sauces can be used to advantage when preparing cephalopods. Because the liquid in the muscle meat of cephalopods seeps out very easily, there is a risk that they will end up tasting very dry, especially if they are cooked over a long period of time or if they were previously frozen. Conversely, the meat can also easily absorb liquid and this is where marinades and sauces come into the picture. It is, therefore, important to be aware of how a particular marinade will affect the muscle meat. The acid in a sour marinade made, for example, with lemon juice or vinegar can denature the proteins in the meat, causing it to become firmer and tougher. Other liquids, such as the cephalopod's own ink, may contain enzymes that help to tenderize the meat.

Sour marinades to which sweet and umami tastes have been added work well with any kind of cephalopod meat, whether raw or prepared in some fashion. This type of marinade has been elevated to a very high level in Japanese cuisine where one finds sixteen traditional versions made with rice vinegars (*su*). Two of these, *ponzu* and *sanbaizu* are especially suitable.

Ponzu, in particular, is an almost magical ingredient in marinades for cephalopods. It is a combination of Japanese soy sauce, *yuzu* juice, rice vinegar, *dashi*, and *mirin*. Its near relative, *sanbaizu* consists of the same ingredients, minus the *yuzu* juice. Bottles of *ponzu* are generally readily available where Asian food products are sold. Both of these sauces keep well under refrigeration.

Sudako-tako is a traditional Japanese dish of marinated octopus. The recipe is a staple of Japanese cuisine and is as simple as it can possibly be. As in other cases, the challenge is cooking the octopus so that it is just tender enough. In his classical book, *Japanese Cooking*, Shizuo Tsuji gives the following instructions: Cut the arms of the octopus into one centimetre slices crosswise. Pour either *ponzu* or a combination of *yuzu* or lemon juice and *sanbaizu* over the slices. Allow the dish to stand for a few minutes or a little longer if it is refrigerated until it is served. Allow four to six slices per person.

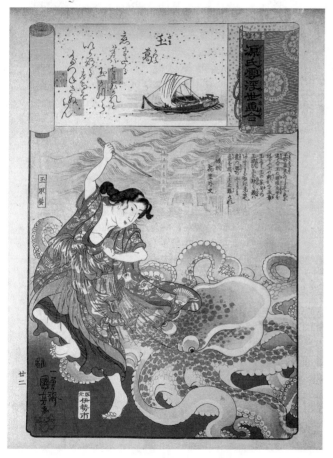

◘ Woodcut by Utagawa Kuniyoshi (1798–1861). The picture shows Princess Tamatori, who is being attacked by, and fighting with, a giant octopus that wants to steal an expensive jewel from her.

Octopus Salad Made the Italian Way

Italian cuisine has a wealth of different, regional recipes for simple octopus salads. Using only the arms from small octopuses or the tips of the arms of large ones results in a more delicate taste.

 Octopus salad

Serves 4

500 g (a little more than 1 lb) cleaned, previously frozen octopus arms

1/2 dL (3 Tbsp) chopped flat leaf parsley

1 clove garlic, finely minced

1/2 stalk celery, sliced thinly

1/2 carrot, sliced into thin strips

1/2 dL (3 Tbsp) extra virgin olive oil

1/2 dL (3 Tbsp) lemon juice

Salt

A little dried oregano

1. Cut the arms into smaller pieces, 3–4 cm (about 1 1/2 in) long, depending on the thickness of the arms.
2. Simmer the pieces in water for 18 minutes and then allow to stand in the water for a while.
3. Drain off the cooking water and reserve for another use.
4. Allow the octopus pieces to cool off and put them in a bowl.
5. Mix together the parsley, garlic, celery, carrot, olive oil, and lemon juice and add to the octopus pieces.
6. Season to taste with salt and oregano.
7. Keep the marinated salad in a cool place for about 1/2 hour before serving.

◘ Octopus salad.

Ceviche

Ceviche may be related to some very old Persian meat dishes prepared with vinegar, examples of which can be found in food cultures in many parts of the world. Above all, it has been embraced wholeheartedly by Peru, which since 2014 has celebrated the 28th of June as National *Ceviche* Day. The Peruvians claim that they have made *ceviche* for at least 2,000 years by steeping fish in lime juice and seasoning it with chili. The recipe for cuttlefish below is inspired by Peruvian *ceviche*.

 Peruvian cuttlefish *ceviche*

Serves 4 as an appetizer
2 Tbsp red wine vinegar
1 Tbsp sugar
1 small red onion, chopped
Salt
Small amount of fresh chili pepper, finely minced
12 small, ripe cherry tomatoes
2 spring onions
Flat leaf parsley
200 g (7 oz) cuttlefish (or squid)
Zest and juice from 3 limes
Olive oil

1. Mix together the red wine vinegar, sugar, chopped red onion, a little salt, and some finely minced chili pepper and set the marinade aside for a time.
2. Cut the tomatoes in half, gently remove the seeds and set aside, slice the spring onions, and coarsely chop the parsley.
3. Cut the cuttlefish up into bite-sized pieces.
4. Pour the marinade over the cuttlefish. Sprinkle with some grated lime zest and squeeze the juice on top.
5. Keep the marinated cuttlefish in a cool place and allow to rest for 15–20 minutes.
6. Arrange the tomatoes, tomato seeds, and parsley in small bowls, and drizzle with a little olive oil.
7. Top with the marinated cuttlefish.

◘ Peruvian cuttlefish *ceviche*.

Squid or Cuttlefish with *Miso*

Miso is a traditional fermented Japanese paste made primarily from soy beans. As it is very rich in umami it is particularly well-suited for marinating and tenderizing vegetables, meat, and fish. The intensity of the resulting taste is dependent on whether one uses white, red, or dark *miso*. To marinate a squid or cuttlefish mantle, cover it entirely with *miso* and keep it in the refrigerator. After one week the mantle will have taken on the colour and taste of the *miso*, its meat will have become firm, and it will have a pleasant texture. The longer the cephalopod is allowed to marinate, the stronger the taste.

To serve, the marinated cephalopod is cut into strips to highlight the contrast between the inner pale part of the mantle and the outer surface that has taken on a dark brown colour. *Miso*-marinated squid or cuttlefish pieces can be added to a green salad or served as a side dish with fish.

Dried Cephalopods

In the fishing ports of mainland Greece and its islands, one can see large octopuses that have been hung up outside to dry so that they can lose as much as one-half of their liquid content before they are cooked on the grill. While the drying process makes the octopuses more tough, they come out crisper when grilled. Fish and cephalopods are dried in a similar way in many parts of Asia.

🔲 Dried octopus arms (*Octopus vulgaris*).

A traditional Japanese breakfast always includes a portion of *ichiya-boshi*, partly dried fish or squid that has been grilled. While drying the fish leaves its flesh quite firm, it also prevents the taste substances from seeping out onto the grill. Typically the fish or cephalopods are prepared by being cleaned at the end of the day and then placed on a wooden rack in a meat safe overnight.

🛈 Grilled semi-dried squid (*ika no ichiya-boshi*)

A piece of mantle from a large squid or cuttlefish, cleaned

Sea salt

Mayonnaise with *yuzu*

Concentrated *dashi*

Shichimi

1. Cut open the cleaned mantle along the side, sprinkle lightly with salt, and place it flat on a rack or a bamboo mat.
2. Allow the mantle to dry in the sun for 2–4 hours until the meat has shrunk together and taken on a

leather-like texture. Alternatively, it can be allowed to dry overnight in a cool place and then placed briefly in the sun in the morning.

3. Grill the mantle for 2–4 minutes on each side on a moderately hot grill, preferably a charcoal grill.
4. Cut the mantle into strips about 1 cm (1/2 in) wide.
5. Place 3–4 strips on each plate, sprinkle lightly with *shichimi*, and add a dollop of the mayonnaise on the side for dipping. If no *shichimi* is available, use a little chili powder instead.

The dried squid or cuttlefish strips could also be served with a little soy sauce or *ponzu*. After step 2, the mantle can be kept in the refrigerator for up to two days before being grilled.

◘ Grilled, semi-dried cuttlefish (*ika no ichiya-boshi*).

In southeast Asia, dried cephalopods are sold as everyday snacks, often found displayed by the check-out counter, much like chocolate bars and candy in western supermarkets.

Those made with Pacific flying squid (*Todarodes pacificus*) are especially popular and come in three varieties, *surume*, *noshi-ika*, and *saki-ika*. *Surume* is the most traditional type, originally made by flattening gutted squid, but otherwise leaving their original shape, and drying them in the sun. This makes them very tough, much like jerky, but they are then grilled lightly and torn into strips. *Surume* is often served with a dipping sauce and enjoyed as a finger food with alcoholic drinks. Some of the softer dried squid are heated and flattened with iron rollers to make very thin sheets, called *noshi-ika*, which may be left as is or seasoned to have a salty, sweet taste. *Noshi-ika* are a simple snack commonly sold at street stalls. *Saki-ika*, on the other hand, are now usually made using a more complex modern process. Gutted whole raw squid are placed for a few minutes in water at a temperature of 65–85 °C. After being cooled, they are shredded into long, thin strands that can be seasoned in a variety of ways. They are then dried for up to twenty-four hours at a temperature of 40–45 °C until they have lost about a half of their liquid content. Next the strands are cured for two weeks and then dehydrated again at 110–120 °C for a few minutes. At this point they have lost so much of their liquid content that they are sufficiently well preserved to keep for a long time in tightly sealed packages. *Saki-ika* takes on a yellowish or brown colour, depending on how it has been seasoned. Like *surume*, it is a favorite bar snack.

Octopuses can be made quite tender by dehydrating the raw arms for one or two hours at 65 °C, depending on their thickness. Sometimes they are first rolled in spices such as curry, after which they can be smoked or grilled and used as a snack. The arms are often cut up into appropriately sized pieces.

◘ Curried dried octopus arms.

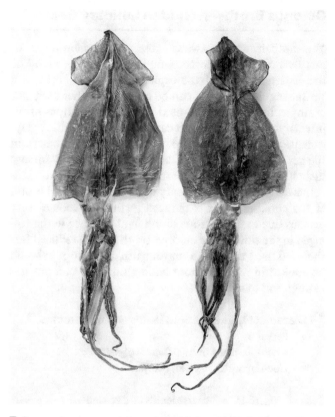

◘ *Surume*: Japanese specialty—whole dried squid (*Todarodes pacificus*).

Octopus Broth—A Hidden Culinary Gem

Yet another surprise awaits those who prepare an octopus by boiling it—the cooking water captures many of the essential tastes of the cephalopod and it becomes a veritable gold mine that can be tapped for other uses. For example, it can simply be used to glaze the octopus arms, intensifying their taste, or it can be made into a thick mayonnaise that can be grilled to resemble *foie gras*, albeit from the sea. The cooking water takes on an especially intense taste if the octopus is prepared in a pressure cooker. It is somewhat sweet, with a meaty taste, much like veal broth. If the cooking water is reduced further it becomes very viscous due to the gelatine found in abundance in the collagen in the skin. If the octopus broth is to be reduced further and used to make a mayonnaise, the octopus should be cooked in water without adding salt. The octopus has enough salt in itself.

ⓘ Glazed octopus arms with lentils and mushrooms

Serves 4

4 octopus arms from *Octopus vulgaris*

Cooking water from an octopus

2 leeks

150 g (5 1/2 oz) coarse lentils of any kind

1 onion, chopped coarsely

2 bay leaves

1 sprig thyme

50 g (1 3/4 oz) small mushrooms (brown beech mushrooms or chanterelles)

25 g (1 oz) butter

Salt and pepper

1. Cover the octopus arms with water, bring to a boil, and simmer for 18 minutes. Remove from heat and allow them to rest in the cooking water for another 18 minutes. Take out the arms and set them aside.
2. Keep the cooking water and reduce it until it thickens.
3. Clean the leeks and slice them crosswise unevenly in pieces that are 2–3 cm (3/4–1 1/2 in) long.

4. Place the pieces in a warm, dry skillet and char them lightly on the cut ends over low heat. Separate the pieces into rings and set them aside.

5. Cook the lentils with the chopped onion, bay leaves, and thyme until the lentils are soft. Drain off any excess water and discard the onion, bay leaves, and thyme.

6. Trim the mushrooms and sauté them in the butter. Season to taste with salt and pepper.

7. Pour a little of the octopus cooking water into the pot with the lentils, cover and simmer for a little while, and season to taste.

8. Simmer the octopus arms in the remaining cooking water until they are glazed.

9. Arrange whole octopus arms on four serving plates with the lentils, mushrooms, and leek slices. Drizzle with the left-over glaze.

◘ Octopus arms glazed with their own cooking water.

ⓘ 'Foie gras from the Sea' on a black brioche with seaweed caramelized in late-harvest dessert wine

Serves 4

Black brioche

500 g (3 1/8 c) flour
20 g (4 tsp) yeast
8 g (1 1/2 tsp) salt
60 g (1/4 c) sugar
60 g (1/4 c) water
5 eggs
1 1/2 tsp squid ink
250 g (8 3/4 oz) butter, at room temperature

'Foie gras from the sea'

1 bottle late harvest dessert wine such as Sauterne
25 g (1 oz) *wakame* or other dried seaweed in small strips or pieces
Cooking water from an octopus (*Octopus vulgaris*)
Neutral tasting oil

The day before

1. Using a mixer with a dough hook, mix together all the brioche ingredients except the butter. Add in pea-sized pieces of butter a few at a time over a period of about 15 minutes until the dough sticks to the hook.

2. Put the dough into a greased baking pan, filling it one-third full, cover with plastic film, and place in a cool spot for 12 hours.

3. Then allow the dough to rise at room temperature for 4–5 hours until it is near the rim of the baking pan.

4. Bake the brioche at 160 °C (325 °F) for 20–25 minutes. Remove from pan and set aside.

5. Pour the wine into a saucepan, add the dried seaweed, and allow it to stand for 5–10 minutes.

6. Simmer until the wine is completely reduced and the seaweed is thoroughly glazed. Dry the seaweed in an oven or a dehydrator at 50 °C (125 °F) until it is crisp.

On the day

1. Reduce the octopus cooking water until it thickens and has a strong taste.
2. Place the reduction in a small bowl and beat with a hand blender while adding the oil a little at a time, as for mayonnaise. (Add oil according to how much one wants to enhance the unique taste of the octopus and how rich the '*foie gras*' should be.)
3. Slice the brioche and toast the slices.
4. Using two spoons, form the '*foie gras*' into egg shapes and place them on the toasted brioche slices. It is possible first to fry the '*foie gras eggs*' on a dry, warm non-stick skillet, turning them carefully.

To serve

Arrange the brioche slices on plates and top with the caramelized seaweed.

◼ '*Foie gras* from the sea' prepared from an emulsion of octopus cooking water, placed on brioche tinted with squid ink and decorated with caramelized *wakame* seaweed.

The biggest surprise about the octopus cooking water is not that it tastes very good, but that it self-emulsifies, that is to say that it will form a mayonnaise-like emulsion just with oil without the addition of an emulsifier such as egg yolk. The gelatine in the cooking water also helps with the consistency and the stability of this emulsion, which can be used as a cold sauce with a variety of cephalopod and shellfish dishes. It can also be prepared as a firm solid that has a texture similar to that of puréed duck or goose liver. This '*foie gras* from the sea' is unusually firm, delightfully creamy, and has an intense octopus taste.

Grilled Cephalopods

One of the simplest, but also most amazing taste experiences can be found along the Spanish and Portuguese coasts if one arrives in a small village around lunch time. The modest restaurants by the beach often serve freshly caught grilled squid or cuttlefish, along with plain boiled potatoes, a small green salad, and a few slices of sun-ripened tomatoes. These dishes are known as *calamares a la plancha* in Spain and *calamares grelhados* and *lulas grelhadas* in Portugal.

Almost invariably these dishes are made with squid. Their innards are removed and they are then placed on the grill whole. They can also be made with large, thick pieces cut from a cuttlefish mantle and grilled quickly on both sides. After just a few minutes they become amazingly tender and crisp on the surface. Sometimes cuttlefish are also grilled whole, but the mantle is scored crosswise. While the cuttlefish keeps its shape, the heat causes the mantle to open up like an accordion. In all cases, it is important to find a balance between the length of time on the grill and how hot it is. The trick is to ensure that the cephalopod meat remains juicy but is still a little crisp, especially the tentacles and arms that should almost be crunchy at their tips.

There is a plethora of recipes for grilling cephalopods. Some call for coating them with oil, lemon juice, and seasonings before cooking, while others drizzle them with a little oil dressing afterwards and serve them with lemon wedges. In the recipes below we have opted for a pair of suggestions for easy, fast ways to grill squid and cuttlefish.

Although the Spanish expression *a la plancha* literally means placed on a flattop metal griddle or a skillet, *calamares a la plancha* are invariably prepared on a grill, preferably over charcoals.

If one wants to present a large cuttlefish so that it still looks whole, one can gut it by making a cut under one of the fins and pulling out the innards and the cuttlebone, leaving the eyes, mouth parts, and so on attached to the mantle.

ⓘ Squid cooked on a grill or over an open fire

Serves 4

1 dL + 1 Tbsp (3/8 c + 1 Tbsp) olive oil

2 organic lemons

Sea salt

1 kg (2 1/4 lb) fresh squid

6 cloves garlic

Flat leaf parsley

1. Make a marinade from 1 dL (3/8 c) of the olive oil, the juice of one lemon, and a little salt.
2. Clean the squid by pulling out the innards through a slit under a fin.
3. Score the mantles on both front and back and place the squid in the marinade.
4. Peel the garlic cloves, slice them lengthwise, and place them in the remaining olive oil on a cold skillet, warming them slowly until the slices turn golden. Drain them on paper towel and sprinkle with salt.
5. Cut the other lemon into wedges and coarsely chop the parsley.
6. Grill the squid, turning them a couple of times. They are ready just when the colour of the meat turns white. Alternatively hang them on a hook over an open fire.
7. Sprinkle with the chopped parsley and garlic slices and serve immediately with lemon wedges.

■ A whole large squid (*Loligo forbesii*) grilled over an open fire.

ℹ Marinated and grilled siphons on lemon grass stalks

Serves 4

4–8 siphons with attached retractor muscles from large squid

Marinade

1 dL (3/8 c) *ponzu*

2 Tbsp fish sauce

1 Tbsp spicy chili sauce

1 tsp sesame oil

2 fresh lime leaves

1 organic lime

1 stalk of lemon grass per siphon

Grilled Cephalopods

1. Chop the lime leaves finely, grate the rind of the lime and squeeze out its juice. Mix with the *ponzu*, fish sauce, chili sauce, and sesame oil to make the marinade.
2. Clean the siphons and marinade them for 30 minutes in half of the marinade.
3. Lightly crush the lemon grass stalks and place the siphons on them just before they are to be grilled.
4. Heat the grill, place the siphons on it, and turn them once.
5. Serve with the remaining marinade.

◘ Marinated, grilled squid siphons with attached retractor muscles on lemon grass stalks.

ⓘ Cephalopod mouth parts and beaks on beans wtih tomatoes and dried shrimp

Serves 4

1 L (32 fl oz) water

70 g (5 Tbsp) salt

Ca. 200 g (7 oz) mouth parts and beaks from all types and sizes of cephalopods

20 small cherry or campari tomatoes

20 g (3/4 oz) black garlic

Small amount of fine salt

2 Tbsp olive oil

1 sprig of lemon thyme

300 g (11 oz) cooked white beans

12 small Brussels sprouts, separated into leaves

2 dL (4/5 c) octopus cooking water, *dashi*, or neutral stock

1/2 dL (3 1/3 Tbsp) dried shrimp

Small amount of white flour

Neutral tasting oil

1. To make the brine, warm the water and mix in the salt. Allow to cool to 5 °C (41 °F).
2. Marinate the mouth parts and beak in the brine for 5 minutes, then remove and set aside.
3. Preheat an oven to 185 °C (365 °F). Bake the tomatoes, black garlic, salt, and olive oil together in a small pan for 6–8 minutes. Remove the tomatoes and black garlic with a slotted spoon and set them aside for later use.
4. Pour the remaining liquid into a pot, bring to a boil with the lemon thyme, add the beans, and allow to simmer for a little while.
5. Heat neutral oil to 165 °C (330 °F) and quickly sauté the dried shrimp in it. Remove the shrimp, place on paper towels to absorb the excess oil, and sprinkle with salt.
6. Toss the mouth parts and beaks in a little flour and fry them in the same oil until they are golden.
7. Just before serving, season the beans to taste and mix in the Brussels sprout leaves.
8. Warm up the tomatoes and black garlic, distribute on plates, add the warm beans with Brussels sprouts, and top with the pan-roasted shrimp, mouth parts, and beaks.

Larger beaks can be too difficult to eat but add an esthetic element to the presentation, which the guests themselves can set aside. The small beaks are a crispy treat and a great delicacy. The muscles from the mouth parts are similar to small meatballs.

◨ Cephalopod mouth parts and beaks on beans with tomatoes and dried shrimp.

ⓘ Sautéed small cuttlefish and potato chips with octopus suckers and mayonnaise

Serves 4

Sautéed cuttlefish

20 small cuttlefish, 4–5 cm (1 1/2–2 in) long, with innards and ink

1/2 L (16 fl oz) olive oil

2 cloves garlic

Sea salt

Potato chips with suckers

1/2 L (16 fl oz) water

12 g (2 1/2 tsp) salt

2 baking potatoes

Suckers from cooked octopus arms or the small tips from the arms

1 egg white, lightly beaten

A few herbs such as fennel, dill, or oregano (optional)

1/2 L (16 fl oz) neutral oil for deep frying

Sautéed cuttlefish

1. Rinse the cuttlefish, dry them well, place them in a dish and pour olive oil over them.
2. Crush the garlic cloves with your hands and sauté them in a skillet in a few spoons of the olive oil from the cuttlefish.
3. Sauté the cuttlefish quickly until golden at a medium heat until they are golden.

Potato chips with octopus suckers

1. Make a brine from the water and salt.
2. Peel the potatoes, cut into very thin slices, and rinse them thoroughly.
3. Heat the brine, pour it over the potato slices. Turn the slices carefully, remove after 5 minutes, and dry off completely with a clean cloth or paper towels.
4. Slice the suckers and/or the tips of the arms into thin pieces.
5. Brush the potato slices on one side with a little beaten egg white. Place the octopus pieces on half of them, as well as a few herbs (optional). Put the remaining half of the potato slices on top and press the two pieces together like a small sandwich.
6. Allow the potatoes to dry off a bit and then fry them in oil at 165 °C (330 °F) until they are golden and crisp.
7. Divide the cuttlefish and the chips into four portions and serve immediately.

To serve

The combination of small cuttlefish that are grilled with their innards and ink sacs and these potato chips might appeal only to die-hard cephalopod lovers. But the chips can be served on their own as an appetizer with homemade mayonnaise seasoned with squid ink and garlic, as a side dish with a little toasted seaweed, or as a garnish for a big cephalopod dinner.

◘ Sautéed small cuttlefish and potato chips with octopus suckers and squid ink mayonnaise.

Deep-Fried Cephalopods

Cephalopods, especially squid and cuttlefish, can be prepared quickly and easily by deep-frying them, sometimes in batter. This does run the danger, however, that they turn out tough and that the batter becomes damp and oily. Badly prepared breaded squid rings are just about the best way to put people off eating cephalopods.

The problem can be solved by taking into account the muscle structure of the cephalopods' mantle. The strong muscles are in a circular arrangement in the mantle. Cutting it into rings crosswise leaves the muscle fibres intact allowing them to become tough when heated. It is much better to slice the mantle along its length, cutting through the strong muscle fibres. So in a way, the best 'squid rings' are actually made from strips.

One way to avoid ending up with a greasy batter is to use *panko* instead of ordinary bread crumbs. *Panko* are a Japanese version of bread crumbs, which are very porous and filled with a huge number of small air pockets causing them to repel the oil, leaving the squid fantastically crisp and crunchy.

◘ Deep-frying squid rings.

ⓘ Extra crisp squid rings and strips

Serves 4

1 L (32 fl oz) water

70 g (5 Tbsp) sea salt

Ca. 500 g (17 oz) cleaned squid mantle plus the arms
and tentacles

Batter

40 g (2 3/4 Tbsp) cornstarch

40 g (2 3/4 Tbsp) high gluten flour

2 pinches of baking powder

Fine salt

Pinch of cayenne pepper

1–1 1/2 dL (2/5–2/3 c) cold soda water

1/2 dL (1/5 c) vodka

1 tsp neutral oil

Ice cubes

Crushed *panko*

1 L (32 fl oz) neutral oil for deep-frying

1. To make a brine, warm the water and mix in the salt.
   Allow to cool to 5 °C (41 °F).
2. Cut the mantle into rings and wide strips 5–10 cm
   (2–4 in) in length.
3. Marinate the pieces in the brine for 5 minutes. Remove and place in a sieve to allow them to dry off.
4. To make the batter, mix together the dry ingredients,
   except the *panko*, in a bowl. Add the soda water, vodka, and 1 tsp neutral oil, and stir carefully to avoid
   making the batter tough. Place the bowl in a bigger
   bowl with ice cubes to keep the batter cool.
5. Heat the oil to 165 °C–175 °C (330 °F–350 °F) and
   place the bowl with the batter and another with the
   panko close by.
6. Dip the cuttlefish pieces in the batter and coat with
   the *panko*. The batter should just stick to the pieces.
   If it is too thick, add a little more soda water. Deep-
   fry until the pieces are golden.
7. Serve with a good, home-made garlic mayonnaise
   and possibly a few lemon wedges.

◼ Deep-fried squid (*Loligo forbesii*) rings and strips, breaded with *panko*.

Pescaíto Frito and Other Mediterranean-Inspired Dishes

One of Europe's oldest cities, Cádiz, in Andalusia, was founded by the Phoenicians around 1100 BCE. In the old part of the town by the flower market at the Plaza de las Flores there is a bar called Freiduria Las Flores I, which serves the most wonderful delicacies from the sea, many made with cephalopods of all kinds. In the Andalusian dialect they are called *pescaito frito*, from the Spanish *pescado frito*, meaning deep-fried fish and shellfish.

Freiduria Las Flores I is a very simple, humble establishment. Customers go right up to the counter to order deep-fried pieces of squid and cuttlefish in a cardboard cone to go or eat them from a plate while standing at the bar or sitting on a stool at one of the small wooden tables. Cold, fresh beer on tap goes down well with the food and one can eat until one is ready to burst. The obvious reason why seats are in such high demand is that the locals know that Las Flores is the best place to eat *pescaíto frito*.

Tapas made from very fresh squid are an important feature of the menu. These are not just the ubiquitous squid rings that are often much too greasy and as tough as rubber. These are much better—the cephalopods have come straight out of the sea and are prepared with just the right amount of oil and breading needed to highlight the taste of the raw ingredients.

Two of the local specialties are *puntillitas* (*calamares chiquititos*) and *calamaritos* (*chipirones*). *Puntillitas* are made from baby squid of the *Alloteuthis subulata* species. *Calamaritos* or *chipirones* are, as their name suggests, small squid of the *Loligo vulgaris* species.

◨ Commercially prepared *chipirones* stuffed with crab meat, readily available in cans in Spain.

Chocos fritos is a nickname for the deep-fried mantle of small cuttlefish (*Sepia officinalis*), which cheekily makes a subtle reference to the Spanish slang word for testicles. A very expensive gastronomic specialty from Andalusia is called *huevos de choco*, but, ironically, it has nothing to do with either eggs (*huevos*) or male reproductive organs. It is made from nidamental glands, which are found in female cuttlefish and are involved in the secretion of the gelatinous substances that binds together the eggs or the cases that protect them. *Huevos chocos* require next to no preparation. They are either grilled or marinated in olive oil, garlic, and parsley.

Another dish, known as *pulpo aliñado* is made up of octopus arms prepared with *aliño* sauce, a mixture of olive oil, wine vinegar (which in Andalusia is, naturally, sherry vinegar), salt, green bell peppers, finely chopped tomatoes, and onion.

Of the many other recipes for octopus, one is so well-known and has such a central place in Spanish cooking that it is almost impossible to miss. We are talking, of course, of *pulpo a la Gallega*, which as the name implies is originally from Galicia. For hundreds of years this region of Spain and especially the area around the port city of Vigo have been a famous centre for octopus fishing. The dish is so completely embedded in the local cuisine that it is also known as *pulpo a feira*, denoting that it is a specialty served on the feast day of the the patron saint of the city of Lugo. The recipe for *pulpo a la Gallega* is extremely simple and calls for only five things: octopus, potatoes, olive oil, paprika, and salt. But this is deceptive, as the trick is to end up with octopus arms that are so buttery soft and tender that sometimes one wishes they had a bit more bite.

🛈 *Pulpo a la Gallega*

Serves 4

1 octopus weighing about 1kg (2 1/4 lb)

12 small potatoes

Sea salt

Olive oil

Paprika

1. Cover the octopus with water, bring to a boil, and simmer for 18 minutes and then let it rest in the water for another 18 minutes.
2. Cook the potatoes whole until they are tender and then peel them. Cut them up into slices, arrange them on a wooden platter, and season with salt.
3. Cut the octopus arms up into pieces, arrange them on the potatoes, drizzle with olive oil, season with paprika, and serve.

◘ The classic Galician dish *pulpo a la Gallega.*

In our own test kitchens we came up with a Nordic variation on this traditional Galician dish. Instead of octopus, it calls for grilled cuttlefish placed on potato purée. We also took the liberty of tinting some of the potato purée with squid (or cuttlefish) ink to make it black and substituting saffron for the paprika. We doubt that this recipe, *calamar a la manera del mar del Norte*, would meet with general approval in Galicia.

Calamar a la manera del mar del Norte

Serves 4

Ca. 500 g (2 1/4 lb) arms and tentacles from a cuttlefish

A few strands of saffron

4 baking potatoes

1 Tbsp squid (or cuttlefish) ink

Sea salt

Olive oil

Crisp squid ink spaghetti (see separate recipe)

1. Cut the arms and tentacles up lengthwise into appropriate sizes.
2. Soak the saffron in 1 Tbsp water.
3. Peel the potatoes, cut them into large chunks, and cook. Drain the water and divide the potatoes between two bowls.
4. Toss the potatoes in one of the bowls with some squid ink and a little salt. Season without turning them too much.
5. Sprinkle the potatoes in the other bowl with a little salt, pour the saffron water over them, and keep the potatoes warm.
6. Sauté the cuttlefish pieces in olive oil in a skillet until they are golden brown and the meat turns white.
7. Push the potatoes with the saffron through a ricer and onto a plate, followed by those with the squid ink. Distribute the cuttlefish pieces on top and garnish with a bit of crisp black spaghetti.

◗ *Calamar a la manera del mar del Norte.*

The simple preparation below shows how easy it is to cook a dish with squid, once it has been cleaned. The dish was developed by the Italian chef Franco Spadaccini at Restaurant La Grotta dei Raselli in Abruzzo, Italy, together with the Danish chef Kasper Styrbæk.

ⓘ Squid with tomatoes and bell peppers

Serves 4

400 g (14 oz) cleaned squid or cuttlefish
Salt for tenderizing
600 g (1 1/3 lb) small tomatoes of different colours
200 g (7 oz) red and green bell peppers
200 g (7 oz) onions
2 cloves garlic
3 Tbsp olive oil
11/2 dL (2/3 c) water
Salt and pepper

1. Cut the squid or cuttlefish meat into thin strips and sprinkle them with salt to tenderize. Set aside.
2. Cut the tomatoes into wedges of different sizes.
3. Trim the peppers, cut them in half, remove the seeds, and cut them into slices crosswise.
4. Peel the onions and slice them thinly.

5. Peel the garlic cloves and press them flat with the heel of your hand. Sauté them in olive oil over low heat for 4–5 minutes.
6. Add the tomato wedges, onion slices, and pepper strips and sauté for another 4–5 minutes at medium-high heat. Add water and allow to simmer for 6–8 minutes. Season with salt and pepper.
7. Transfer the vegetable mixture to an attractive oven-proof dish and distribute the cephalopod pieces on top.
8. Bake in a pre-heated oven at 220 °C (425 °F) for 10–12 minutes.
9. Serve with slices of crusty country-style bread.

▣ Squid with tomatoes and bell peppers.

ⓘ Baked octopus arms with sweet potatoes, chipotle, and lovage

Serves 4

4 octopus-arms, ca. 600–700 g (1 1/3–1 1/2 lb)
1 kg (2 1/4 lb) sweet potatoes
2 Tbsp olive oil
Salt and pepper

Marinade

8 Tbsp ketchup
4 Tbsp soy sauce
2 Tbsp Worcestershire sauce
1 Tbsp rice vinegar
2 Tbsp muscovado or demerara sugar
1 clove garlic, peeled and put through a press
Juice from 2 limes
1/4 tsp chipotle pepper
Salt and pepper

Lovage *gremolata*

1 bunch spring onions (about 6–8)
2 organic lemons
1 clove garlic, peeled and put through a press
6 stems of lovage, chopped finely

1. Cover the octopus arms with water, bring to a boil, and simmer for 18–20 minutes. Turn off the heat, and leave them in the cooking water for a further 18–20 minutes.
2. Mix together all the ingredients for the marinade.
3. Place the octopus arms in an oven-proof pan, toss with the marinade, and leave them for at least 1 hour.
4. To make the *gremolata*: Cut the spring onions into thin slices and mix carefully with the other ingredients. Leave in a cool place until ready for use.
5. Scrub and trim the sweet potatoes, cut them into wedges lengthwise, toss with the olive oil, and season with salt and pepper. Bake in a pre-heated oven at 210 °C (410 °F) for 15–20 minutes. Set aside.
6. Bake the octopus arms in a pre-heated oven at 185 °C (365 °F) for 15–20 minutes.
7. Put the sweet potatoes back in the oven for the last few minutes to reheat them.

To serve

Arrange the octopus arms and potatoes on a serving dish and sprinkle the *gremolata* on top.

▣ Baked octopus arms with sweet potatoes, chipotle, and lovage.

Steamed and Cooked Cephalopods

There is a long list of classic recipes from around the world in which cephalopods are prepared using heat, either by steaming, simmering, or cooking them in water. Most of these are for octopuses, but in the one below we steam cuttlefish instead.

ⓘ Steamed cuttlefish with spinach, fish roe, and a creamy sauce

Serves 4

Ca. 400 g (14 oz) cleaned cuttlefish or squid mantle

2 shallots

100 g (3 1/2 oz) button mushrooms

2 dL (4/5 c) white wine

5 peppercorns

1 bay leaf

Cuttlefish off-cuts (optional)

2 1/2 dL (1 c) whipping cream

Salt and freshly ground pepper

200 g (7 oz) fresh spinach

2 spring onions

1 L (32 fl oz) water

30 g (1 oz) sea salt

A few sprigs of parsley (optional)

One or two stalks of celery

4 Tbsp fish roe, either lumpfish or possibly caviar for a special occasion

25 g (1 2/3 Tbsp) butter

1. Dry off the mantle. Using a sharp knife score the inside in a diamond pattern 2–3 mm (1/10 in) in width.
2. Cut the mantle into 12 squares 4–5 cm (1 1/2–2 in) in size and set aside.
3. Peel the shallots, trim the mushrooms, and chop both.
4. Pour the white wine into a pot, add the onions, mushrooms, peppercorns, bay leaf, and cuttlefish off-cuts (optional). Simmer and reduce the liquid to approximately one-third.
5. Pass through a sieve, return the liquid to the pot, add the cream, simmer a little while, and season with the salt and pepper. Set aside.
6. Trim and wash the spinach and chop up the white parts of the spring onions.
7. Combine the water, sea salt, parsley, and celery to make a brine and bring to a boil. Add the mantle pieces and let them cook for 1–2 minutes. The pieces will curl up as they cook. Remove the pieces from the cooking brine and roll them in the fish roe.
8. Wilt the spinach in the butter in a skillet and season well.
9. Whip the cream sauce a little with a hand blender so that it is foamy.
10. Arrange the spinach on plates, top with the cuttlefish, and distribute the sauce around the pieces.
11. Sprinkle with a few rings cut from the green stems of the spring onions. The dish can also be decorated with a little parsley or cilantro.

◨ Steamed Cooked cuttlefish with spinach, fish roe, and a creamy sauce.

Wok-Cooked Cephalopods

炒鱿鱼 (chǎo yóuyú) is the Chinese expression for deep-fried squid but it is also used as a way to say that someone is being fired from a job. The connection goes back to China in times past when servants brought their own mattress with them when they were hired. Then when they left or were let go they were told to roll up the mattress, which then looked much like a whole deep-fried squid.

Sautéing squid in a little oil in a wok is an ideal way to prepare them because it is possible use high heat for a short period of time, tenderizing the squid while also preserving their juices. An added advantage of using a wok is that the sloped sides allow one to prepare a complete dish in one pan by adding the ingredients in sequence so that all of them end up cooked just right at the same time. For example, one such combination would be squid with a variety of vegetables and mushrooms.

One of the most traditional Cantonese dishes is *jiaoyán yóuyú*, salt and pepper squid. Another easy wok recipe is *zajin chao xianyou*, deep fried squid with sugar peas. The squid are coated with a mixture that leaves them perfectly crisp—semolina flour adds firmness, cornstarch results in crispness, and white flour holds it all together.

ⓘ Wok-fried salt and pepper squid

Serves 4

Ca. 300 g (11 oz) squid, pen and innards removed
1 Tbsp Chinese rice wine (or *mirin* or dry sherry)
1/2 tsp sesame oil
1 1/5 L (5 c) plus 1 Tbsp neutral tasting oil
3/4 dL (1/3 c) semolina flour
3/4 dL (1/3 c) cornstarch
3/4 dL (1/3 c) white flour
1 tsp salt
1/2 tsp white pepper
2 tsp fresh ginger, chopped finely
5 cloves garlic, cut into thin slices
2 hot green chilis, cut into thin strips

1. Cut the heads and arms off the squid, leaving them in one piece. Cut the bodies into rings.
2. Dry off the squid and place them in a bowl.
3. Pour the wine and sesame oil into the bowl and stir it to moisten the squid.
4. Pour 1 1/5 L (40 fl oz) oil into a deep saucepan, keeping 1 Tbsp aside, and heat it to 160 °C (320 °F).
5. Meanwhile, in another bowl mix together the semolina flour, cornstarch, white flour, salt, and white pepper.
6. Coat the squid pieces with the mixture, one handful at a time.
7. Using a slotted spoon, immerse the squid rings in the hot oil, moving them back and forth until they are golden. Fry only a few at a time so that the temperature of the oil remains more or less the same.
8. Place the fried squid rings on paper towels on a plate to draw off excess oil.
9. Taste one of the rings to see if they are sufficiently seasoned with salt and pepper. If not, add some to the flour mixture before frying the next batch. Continue frying the squid in small batches until all are ready.
10. Pour 1 Tbsp oil into a wok at medium heat, add the ginger, and shortly after the garlic.
11. When the garlic has turned golden add the chili strips and sauté them for about 30 seconds.
12. Add the squid to the wok and sauté very rapidly for about a minute mixing the rings with the other ingredients.
13. Serve at once, with cooked white rice if desired.

■ Salt and pepper squid cooked in a wok with green chilis.

ℹ Cephalopod 'pine cones' with sugar peas

Serves 4

Ca. 250 g (9 oz) squid or cuttlefish mantles
200 g (7 oz) sugar peas
Neutral tasting oil
2 tsp oyster sauce
1 tsp salt
2 tsp Chinese rice wine (or *mirin* or dry sherry)
4 Tbsp water
30 g (2 Tbsp) finely chopped fresh ginger
1 tsp finely chopped garlic
1 tsp cornstarch

1. Dry off the mantles. Using a sharp knife score the insides in a diamond pattern 2–3 mm (1/10 in) in width.
2. Cut the mantle into 12 squares 4–5 cm (1 1/2–2 in) in size.
3. Blanch the pieces in boiling water for a short time so that they curl up like a tube and then dry them off.
4. Trim the sugar peas, blanch them in warm oil, and allow them to drain on paper towels.
5. Mix together the oyster sauce, salt, rice wine, and water and set aside.

6. Warm up the wok, add 1 Tbsp oil and sauté the ginger and garlic in it.

7. Add the cephalopod tubes and the sugar peas and pour the oyster sauce mixture over them. Bring to a boil and add the cornstarch dissolved in a little water to thicken.

8. Serve at once.

◘ Pieces of cuttlefish that have been scored on one surface and then cooked in a wok with sugar peas. The pieces take on the appearance of pine cones.

Stuffed Squid and Cuttlefish

Going right back to ancient times, dishes made with stuffed squid have been a feature of traditional Italian cuisine. Many different ingredients are used to fill the squid tubes—spinach, mussels, crab meat, and smoked ham, to name just a few. Sometimes the stuffing is even made from the arms, tentacles, and fins. The first recipe below draws its inspiration from the way in which such a dish is made in the Abruzzo region. It is also possible to make 'inside out' stuffed squid by putting the pieces in a hollowed out red pepper. These look a little like Santa Claus hats, as in the second recipe.

ℹ️ Stuffed squid Abruzzo style

Serves 4

4 whole squid (preferably *Loligo forbesii*) with 10–12 cm (4–5 in) long mantles

Stuffing

75 g (3/4 c) *panko* (or breadcrumbs)

1–1 1/2 dL (2/5–2/3 c) milk

Squid arms and tentacles

60 g (4 Tbsp) grated Parmigiano Reggiano

2 eggs

Salt

Pepper

Tomato sauce

50 g (1 3/4 oz) shallots, peeled and trimmed

2–3 cloves garlic, peeled

2 Tbsp olive oil

50 g (1 3/4 oz) red bell pepper

400 g (14 fl oz) can peeled tomatoes

1 dL (2/5 c) unpitted olives

1/2 dL (3 1/3 Tbsp) capers

Stuffing

1. Clean the squid, remove the innards and pens, cut off the arms and the tentacles. Set aside the tubes and chop the arms and the tentacles coarsely.
2. Soak the *panko* in the milk for 5 minutes and then put it in a sieve to drain off any milk that has not been absorbed.
3. Mix together the chopped squid pieces, *panko*, Parmigiano Reggiano, and eggs.
4. Season to taste with salt and pepper and put the stuffing into the squid tubes.

Sauce

1. Finely chop the shallots and garlic cloves. Place in a skillet with the olive oil and gently cook the mixture until the shallots are transparent.
2. Clean the bell pepper and chop it coarsely. Mix it with the tomatoes, olives, and capers and add all to the skillet.
3. Heat the sauce to the boiling point, lay the squid tubes in the sauce and simmer them for about 30 minutes, turning them a couple of times.
4. If needed, add a little water to the skillet to prevent the sauce from cooking away completely.

To serve

Divide the squid tubes and sauce among four serving bowls. Bake at 185 °C (365 °F) for 25 minutes. Serve immediately.

🛈 'Santa Claus hats'— small sweet peppers stuffed with squid

Serves 4

8 small squid (or cuttlefish or octopuses)

1 L (32 fl oz) warm water

70 g (4 2/3 Tbsp) salt

2 g (2/5 tsp) dried hibiscus flowers

200 g (7 oz) deboned, skinned white fish

Salt

1 piece of chili pepper according to taste, chopped finely

A little leaf parsley, chopped finely

1 egg

100 g (3 1/2 oz) mascarpone

20–30 g (1 1/3–2 Tbsp) breadcrumbs

Freshly ground pepper

400 g (14 fl oz) can roasted *piquillo* peppers (or other small sweet peppers)

40 g (2 2/3 Tbsp) capers

2 dL (4/5 c) neutral tasting oil for frying

150 g (2/3 c) butter

2 shallots, finely chopped

4 Tbsp olive oil

1. Rinse and clean the cephalopods, remove the innards, and cut off the arms and the tentacles keeping them in one piece.
2. Slice the mantles and fins into strips lengthwise and then into smaller pieces.
3. Make a brine by dissolving the salt in the warm water in a saucepan and cool the mixture to 5 °C (41 °F).

4. Place the cut-up squid pieces and the arms and tentacles in the brine for 7 minutes. Remove and allow to drain in a sieve.

5. Crush the dried hibiscus flowers with a mortar and pestle.

6. Mince the fish finely in a food processor. Mix in a little salt, the chili pepper, half the parsley, and the crushed hibiscus.

7. Transfer the fish mixture to a bowl and add the egg and mascarpone. Mix in the squid pieces. Bind the mixture with sufficient breadcrumbs to hold the stuffing together. Season with salt and pepper.

8. Pat the *piquillo* peppers dry, season with salt and pepper, and stuff with the squid mixture.

9. Fasten an arm/tentacle piece to the stuffing of each pepper so that they stick out at the top. Refrigerate until needed.

10. Drain the capers and dry on a paper towel. Heat the neutral oil in a small saucepan to 165 °C (330 °F) and sauté the capers in it for a few seconds.

11. Remove the capers with a slotted spoon and place on a paper towel to absorb excess oil.

12. Place the stuffed peppers in an oven pan, drizzle with the olive oil, and bake at 185 °C (365 °F) for 10–15 minutes, depending on their size.

13. Brown the butter in a saucepan and then add the remaining parsley and shallots.

14. Just before serving pour the melted butter over the squid 'Santa Claus hats' and sprinkle with the crisp capers.

Stuffed Squid and Cuttlefish

◘ 'Santa Claus hats'—small sweet peppers stuffed with squid.

Smoked Cephalopods

All the various types of cephalopods can be cold-smoked very successfully, after which they will keep under refrigeration for a long time. Once smoked, they can also be cut into portions and frozen for later use. Smoked cephalopod meat is versatile and can be incorporated into a whole range of dishes. It can also be substituted for fresh, cooked, or steamed cephalopods in many of the preceding recipes.

If one has access to a cold smoker, it is possible to prepare octopuses, squid, and cuttlefish by hanging them in the smoker overnight. Alternatively, octopus arms can be cooked first and then smoked for a short period of time until they have taken on the desired smoky taste.

Slices of cold-smoked octopus arms are a wonderful addition to plain mashed potatoes, a simple omelette, or a vegetable soup. Strips of the mantle from smoked cuttlefish can be used to add interest to a green salad. If fins from large *Loligo* squid are cold-smoked overnight they become hard and dry. They can then be planed or grated and sprinkled on a finished dish, much like the Japanese *katsuobushi*.

◘ Potato soup with slices of cold-smoked octopus (*Octopus vulgaris*).

◨ Cold-smoked whole squid (*Loligo forbesii*).

Inky Dishes and Snacks

The dark liquid found in the ink sacs of cuttlefish and squid can be used to dye starchy foods such as thickened sauces, bread, pasta, and rice. The resulting deep black colour creates an elegant contrast in dishes with vegetables and, even more so, if these dark foods are paired with the light and white flesh of cephalopod arms and mantles.

🛈 **Black potato *gnocchi* with squid**

Serves 4

Black potato gnocchi

300 g (10 oz) potatoes

1 egg

1 tsp squid or cuttlefish ink

Pepper

1 tsp salt
Ca. 125 g (3/4 c) white flour
1 Tbsp olive oil
Gnocchi with squid
200 g (7 oz) cleaned squid
60 g (4 Tbsp) butter
1 portion *gnocchi*
A small amount of oregano, dried or finely chopped

Black potato gnocchi

1. Peel the potatoes and cook them until they are tender.
2. Drain off the water and steam the potatoes slightly so that they will be dry. This should yield about 250 g (9 oz) of cooked potatoes.
3. Rice the potatoes. Put them on a floured pastry board and make a hollow in the middle.
4. Put the egg, ink, pepper, and 1 tsp salt in the hollow. Fold the mixture together and make a new hollow.
5. Add the flour to the hollow and mix everything carefully. Knead as little as possible, just until the mixture is uniform.
6. Sprinkle flour on a smooth surface (counter top or pastry board). Working with a quarter of the mix at a time, roll it out into a long cylinder, about 1 1/2 cm (a little more than 1/2 in) thick. Cut the roll into small pieces, about 1 cm (1/3 in) each.
7. Bring a pot of lightly salted water to a boil. Add the *gnocchi* in batches, cooking until they float to the surface. Remove the *gnocchi* with a slotted spoon and transfer them immediately to ice cold water.
8. Strain the cooking water and reserve some of it for making the sauce.
9. Drain the *gnocchi* and coat them with a little olive oil so that they do not stick together.
10. Store the *gnocchi* in a cool place until making the rest of the dish. They can be made a day in advance.

Gnocchi with squid

1. Cut the squid up into smaller pieces, melt the butter in a skillet, and sauté the squid very quickly.
2. Add the *gnocchi* to the skillet to warm them through, season with oregano to taste, and add 2 Tbsp of the cooking water to make a slightly creamy butter sauce.
3. Season with salt and pepper to taste and serve in pasta bowls.

◘ Black potato *gnocchi* made with cuttlefish ink, served with crisp pieces of squid (*Loligo forbesii*). Arms, tentacles, strips of the mantle, and the siphon are all used.

Black spaghetti with squid

Serves 4

Homemade black spaghetti

400 g (2 1/2 c) farino 00 or all-purpose flour

4 eggs

2 tsp squid or cuttlefish ink

Squid

300 g (10 oz) squid, both mantle and arms

2–3 cloves garlic

3/4 dL (1/3 c) olive oil

1 portion black spaghetti

1/2–1 dL (3–6 Tbsp) cooking water from the spaghetti

Black spaghetti

1. Knead all the dough ingredients together. Roll out in sheets so that they are appropriate sizes for putting through a pasta maker at the level 2 thickness.
2. Feed through the pasta maker at the spaghetti setting.

Squid

1. Cut the squid mantles up into strips that resemble fettucine and separate the arms from each other.
2. Crush the cloves of garlic with the heel of your hand, be place them in a skillet with the olive oil. Sauté at medium heat until the garlic is golden.
3. Add the squid pieces and simmer for about 5 minutes. Turn off the heat.
4. In the meanwhile, bring a pot of salted water to a boil and cook the spaghetti for about 2 minutes. (If using ready-made pasta, follow the directions on the package, but reduce the time a little.)
5. When the pasta is almost *al dente*, drain it in a colander, reserving a small amount of the cooking water. Add the spaghetti to the squid in the skillet.
6. Turn the heat up to medium, add a little of the cooking water so that the dish becomes creamy. Simmer until the spaghetti is *al dente*.

◧ Squid ink pasta in a dish with grilled squid arms and tentacles.

ⓘ Black hot dog buns filled with seaweed flavoured onions

Serves 4

Hot dog buns

2 1/2 dL (1 c) water

12 1/2 g (2/5 oz) cake yeast or 1 package (2 1/4 tsp) dry yeast

35 g (2 1/3 Tbsp) sugar

Ca. 475 g (3 c) flour

1 egg

1 Tbsp squid or cuttlefish ink

10 g (2 1/3 tsp) sea salt

35 g (2 1/3 Tbsp) butter at room temperature
Strong coffee for glazing
Maldon sea salt

Onion and seaweed filling
1 sheet *nori* seaweed
200 g (7 oz) yellow onions
10 g (2 1/3 tsp) butter
2 Tbsp rice vinegar
Salt

Hot dog buns
1. Warm the water to 30 °C (85 °F). Stir in the yeast and sugar. Allow the yeast to become very active.
2. Mix in half of the flour, a little at a time, followed by the egg, ink, salt, and butter. Knead in the rest of the flour until the dough is uniform and no longer sticks to the hands.
3. Allow the dough to rise for about 20 minutes.
4. Divide the dough into 16 portions, shape like hot dog buns, and place them on greased baking sheets. Cover with a damp cloth and allow them to rise until they have doubled in size.
5. Brush with strong coffee, sprinkle with some Maldon salt, and bake at 170 °C (340 °F) for 10 minutes or until done.

Seaweed flavoured onions
1. Soften the sheet of *nori* in water and then cut it into strips.
2. Cut the onions in half and then slice thinly.
3. Melt the butter in a saucepan, add the onion slices and allow them to simmer until they are soft. Add the seaweed strips and simmer for a further 5 minutes.
4. Season to taste with rice vinegar and salt.

Filling
Another approach is to cut the buns in half and top them with an assortment of ready-to-eat cephalopod pieces cut into suitable sizes. This is a good way to use up leftover bits such as siphons, liver, and mouth parts.

🔳 A new take on hot dogs. Black buns made with cuttlefish ink, cut in half and topped with an assortment of prepared cephalopod pieces.

ⓘ Crisp spaghetti with crushed sunflower seeds

Serves 4

25 g (1 oz) sunflower seeds

2 Tbsp oilve oil

3 Tbsp soy sauce

Black spaghetti (see recipe above)

Neutral tasting oil for frying

1. Toast the sunflower seeds in the olive oil in a skillet until they puff up. Mix in the soy sauce.
2. Place the seeds on a paper towel to absorb extra fat and allow them to cool. Then crush them in a mortar.
3. Toss four portions of black pasta in the crushed sunflower seeds. Place the pasta on baking paper and allow it to dry out at room temperature for about 10 hours.
4. Deep-fry the spaghetti until crisp in neutral tasting oil heated to 165 °C (325 °F).

Serve as a snack or with *Calamar a la manera del manera del mar del Norte* (see recipe above).

ⓘ Squid ink *pasta fritta*

Ca. 40 pieces

1 1/2 dL (2/3 c) milk

16 g (1/2 oz) cake yeast or 1 package (2 1/4 tsp) dry yeast

30 g (2 Tbsp) butter at room temperature

1 tsp salt

2 pinches of cayenne pepper

30 g (2 Tbsp) finely grated Parmigiano Regianno

1 1/2 tsp squid or cuttlefish ink

5 g (1 tsp) finely chopped fresh rosemary

300 g (ca. 2 c) white flour

1 L (32 fl oz) neutral tasting oil for frying

Maldon sea salt

1. Warm the milk to 30 °C (85 °F). Stir in the yeast and allow it to become very active. Mix in the melted butter, salt, cayenne pepper, grated cheese, ink, and rosemary.
2. Place the flour on a pastry board, make a hollow in the middle, pour in the liquid and work it outward to make a dough.
3. Cover the dough and allow it to rest for 30 minutes.
4. Roll out the dough to a thickness of 3 mm (1/10 in). Using a small cookie cutter, cut into rounds and sprinkle with Maldon sea salt (optional). Deep-fry in small batches in oil at 165 °C (325 °F) until they puff up and are crisp.
5. Remove with a slotted spoon and place on paper towels to absorb excess oil.

◘ Squid ink *pasta fritta* and crisp black spaghetti with crushed sunflower seeds. These can be used as a snack or as a garnish.

ⓘ Sweet *azuki* beans in squid ink

Serves 4
400 g (2 c) sweet red *azuki* beans
4 dL (1 2/3 c) water
2 g (1/2 tsp) agar
Cuttlefish ink

1. If using canned *azuki* beans, rinse them in the water, place them in a sieve to drain while collecting the rinse water.
2. If using dried beans, cook according to directions. Drain over a bowl, reserving the cooking water in a small saucepan. There should be about 4 dL (1 2/3 c) of liquid.
3. Line a small mold with something that will withstand heat, for example aluminum foil or baking paper. Distribute the beans in the mold.
4. Add a little cuttlefish ink to the reserved water and bring the liquid to a boil. Add the agar while stirring vigorously until it is completely dissolved.
5. Pour the hot liquid into the mold. Stir everything carefully a few times as the agar is setting to prevent all the beans from sinking to the bottom.
6. Place the mold in the refrigerator to cool for about 30 minutes.
7. Unmold and slice into small pieces that are served as a sweet.

Azuki beans are so sweet that they can be used as filling for cakes and in jellied desserts without the addition of any sugar. In Japan *azuki* bean paste is stiffened with agar to make a popular dessert called *yōkan*, often coloured green with *maccha* tea. This recipe is basically a black version of *yōkan*.

◧ A sweet made from red *azuki* beans in agar jelly that is coloured black with cuttlefish ink.

◘ Stuffed squid (*Loligo forbesii*), here in a dish inspired by a recipe for stuffed pike dating from 1653.

Long-Finned Squid in the Style of *The Compleat Angler*

The world's most famous book for sports fishers is *The Compleat Angler* by the British writer Izaac Walton (1593–1683), which was first published in 1653. The story unfolds over a period of five days, as reflected in the five chapters of the book, which consist of a dialogue between two anglers about how to catch and prepare fish.

The fourth chapter is devoted to a discourse about pike, a fish known as "the Tyrant of the Rivers," and how it should be caught using a frog as bait. The gutted fish is enclosed in a 'cage' made from willow branches and grilled whole according to a wonderful, simple recipe that calls for a lot of oysters stuffed into the body cavity. Walton also suggests that anchovies should be part of the stuffing and in the sauce. Altogether these are the basis for a dish with an abundance of savoury umami tastes.

We have adapted Walton's classic recipe to prepare a large long-finned squid (*Loligo forbesii*), stuffed with oysters, crab meat, bread crumbs, and anchovies. The innards of the squid are extracted through a slit in the mantle under one of the fins, leaving the mantle, the part with the eyes, the siphon, mouth parts, and arms and tentacles in one piece. The stuffing replaces the innards and the squid is then placed in the 'cage' woven from willow branches that are so green that they can survive the heat of the grill.

One should all try this recipe at least once in a lifetime!

"This dish of meat is too good for any but Anglers or very honest men [and women]: and I trust, you will prove both, and therefore I have trusted you with this secret."

Izaac Walton
The Compleat Angler (1653)

Grilled squid in the style of *The Compleat Angler*

Serves 4

Willow branches for making a 'cage'

1 very large long-finned squid (*Loligo forbesii*),
 weighing about 2 kg (4 ½ lb)

300 g (10 1/2 oz) crabmeat

Freshly ground pepper

2 eggs

10 fresh oysters

3 anchovy fillets in oil

10 g (2 tsp) savory chopped coarsely

30–40 g (1/4–1/3 c) breadcrumbs

It might be a good idea to co-opt someone who is into crafts to make the 'cage.'

1. Soak the willow branches in water for three days.
2. On the day itself: First measure the length of the squid as it is important that the grill 'cage' will be long enough to encase the squid completely. Bind the willow twigs together—out of respect for the squid, try to make the knots decorative.
3. Make a small slit under of the fins of the squid and pull out the innards and the squid pen. Rinse out the cavity as necessary.
4. Chop one half of the crabmeat in a food processor, mix with freshly ground pepper and the eggs, and place the mixture in a bowl.
5. Shuck and clean the oysters, reserving their liquor and then add it all to the stuffing mixture.
6. Chop the anchovies and stir them into the stuffing mixture together with the chopped savory and the remaining crabmeat.
7. Stir in enough breadcrumbs so that the stuffing mixture will stick together. As both the oysters and the anchovies are quite salty, there is no need to add any salt.
8. Carefully fill the stuffing into the squid's cavity through the slit in its side. Close the slit with a wooden skewer or two.
9. Place the squid in the 'cage,' possibly allowing the arms to stick out a bit.
10. To cook the squid, place the 'cage' on a grill, over an open fire, or in a wood-fired oven. The key aspect is that the core temperature should be 44 °C (110 °F), which will take approximately 25–40 minutes, depending on the size of the squid.
11. To serve, remove the skewers holding the 'cage' closed, carefully take out the squid and share it out into four portions.

Long-Finned Squid in the Style of *The Compleat Angler*

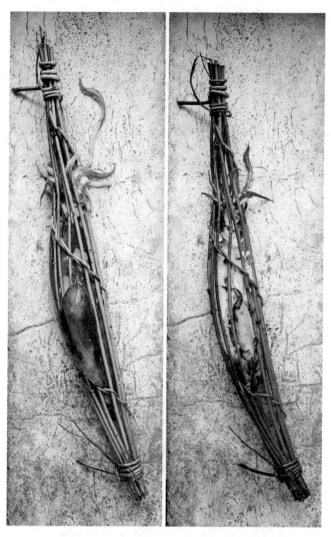

◘ Grilled squid (*Loligo forbesii*) in the style of *The Compleat Angler*, both before it has been cooked (left) and afterward (right).

Cephalopods in Vietnamese and Cambodian Cuisine
One of us (Klavs), together with his wife Pia had the opportunity to make an extended trip to Vietnam and Cambodia to hone his culinary skills, travelling with a backpack, an open mind, keen senses, and an empty stomach!

We started off in the southernmost part of Vietnam at the fish market in Duong Dong on the island of Pho Quoc. Next we visited Cambodia and the crab and cephalopod market in the old French colonial city Kep, with its grand, elegant buildings and broad boulevards, reminiscent of Paris. The third stop was at the pepper farms in the Kampot region and the deserted part of the jungle island Koh Rong.

The voyage did much to enhance my appreciation of the incredibly welcoming and friendly people of this part of the world and the fantastic raw ingredients that provided a livelihood for these very poor fishers and their industrious spouses.

ⓘ Asian squid at its very best

Serves 4
8 small Japanese flying squid (*Todarodes pacificus*)
8 bamboo skewers
8 vines of fresh green peppercorns
1/2 dL (3 1/3 Tbsp) fish sauce
1/2 dL (3 1/3 Tbsp) lime juice
Sugar to taste
1 large clove garlic, crushed
A little fresh chili pepper, finely chopped
Sea salt, to taste
A little lime juice
1 fresh red chili pepper, seeded and finely chopped
1 ripe mango
1 green mango
A variety of herbs to add character

1. Using your fingers or tweezers very carefully remove the innards from the squid.
2. Skewer the squid from their beaks to the tip of their mantles.

3. Make 3 or 4 slits in the top side of the mantle of each squid until the knife hits the skewer.
4. Weave the green peppercorn vines through the slits. Refrigerate until ready to use.
5. Mix together the fish sauce, lime juice, sugar, garlic, and fresh chili to make a sauce. Set aside for 30 minutes to allow the flavours to blend.
6. To make the chili salt, mix together the salt, a little lime juice, and the red chilli. Place in a small container and refrigerate.
7. Peel the ripe mango, remove the pit, cut up into uniform pieces, and drizzle with a little lime juice. Peel the green mango, remove the pit, and cut up into strips.
8. Grill the squid or sauté them in a skillet at high temperature for 4–8 minutes.

Serve with the two types of mango, sprinkled according to taste with the chili salt and chopped up herbs, if desired. Dip the squid in the sauce and enjoy.

◘ Grilled Japanese flying squid (*Todarodes pacificus*) with ripe and green mango, stalks of fresh green peppercorns and a sauce made with fish sauce, lime juice, garlic, sugar, chili, a sprinkle of chili salt, and *rau dang* (a Vietnamese herb).

On a hot day—it was 32°C—we had an appointment for one o'clock to meet Cuong Pham, the director and owner of what is considered the best fish sauce factory on Phu Quoc Island. Our arrangement was intriguingly veiled in mystery as we were given no address but just told to make ourselves ready and stand in the middle of Nguyén Trung Truc Bridge at the mouth of the harbour.

The Duong Dong River divides the town in half. On the Duong Dong side there is a night market. On the other side one finds the colourful local fish market where just about all the goods are not just fresh, but still alive. As is common in Asia, fish, shellfish, and molluscs are kept in aerated basins. Among them were an enormous number of squid that were sending out brightly coloured flashes of light vying for attention, in competition with the fishers' wives who, for their part, were trying to attract the attention of all of us hungry visitors.

We were ready at the appointed time and were looking around in all directions when we suddenly spotted a large flat-bottomed fishing barge sailing toward us. Cuong Pham, who was wearing Ray-Ban sunglasses and a Red Boat t-shirt, was standing in the prow in front of about ten tons of sea salt. As there was a full moon, when the night is not dark enough for fishing, the canal was packed with boats that were staying in harbour. He pointed to three or four boats that we could hop across to get on board his.

We immediately struck up a conversation with the very amiable Cuong Pham, who actually lives in California, which is the biggest market for his fish sauce and where his business is now based. He trained as an engineer and left Vietnam in the 1970s, smuggled out on a boat, which is how his brand of fish sauce came to be named Red Boat.

Cuong's encounters in the United States with poor quality fish sauce led him back to find out more about his uncle's fish sauce, first to rediscover it for his own use and eventually as a way to make a living. According to him, the secret behind his product depends on three things—the black anchovies found in the waters near the island, the kind of salt used to cure the anchovies together with starting the process on the boat only minutes after the fish are caught, and making sure that they start to dry out right away. Over and above that there are very particular climatic conditions on Phu Quoc that are ideal for fermenting the fish.

The anchovies are cured in 185 special barrels made from three different types of wood from the southern part of Vietnam. Each of them can hold fourteen tons of the salted fish. Every month between ten and fifteen barrels' worth are ready to move on to the next stage, fermentation for a whole year. During the last month of fermentation, the bottom layer of the sauce is poured over the top so that it filters itself, leaving a uniform, beautiful golden liquid. This fish sauce is then shipped to the United States where it is bottled and distributed.

We were given the opportunity to do a taste comparison between sauces that were at the three month, seven month, and almost finished stages. The taste impressions ranged from the somewhat raw, but palatable, to full-bodied with notes of Parmesan cheese, to a finished, amber-coloured, smooth, and elegant fish sauce that was very viscous and just cried out to be used in the kitchen.

Our curiosity was rewarded when we enquired whether any fish sauce is made exclusively from Vietnamese cephalopods. Cuong proudly pointed to the little row of clay vats that stood along the edge of the river. He explained that two years ago, as an experiment, he had started to ferment small, local squid using just salt. We sampled the still raw, fermenting sauce and it was like tasting the innermost essence of all that is best about a squid. It had a unique, soft and clean taste gathered into small drops that were abundant in umami. Cuong is deliberating whether it will turn out to be too expensive to produce on a commercial basis. But I was willing to attest that every serious chef would give an arm and a leg for a bottle. I certainly hoped that our small water bottle filled with the sauce would make it safely back to Denmark.

Another day we visited the famous crab market in Kep by Chhak Kep Bay very early in the morning. Most of the products for sale are specialties such as blue swimming crabs (*Portunus pelagicus*) and a variety of octopuses, squid, and cuttlefish, all brought on shore around dawn. The fishers' wives await their arrival and set to work sorting out the catch and dividing it up. The crabs are placed in woven baskets and put back in the water, where they can be viewed so that one can select some to eat—the quantity and size depending on one's appetite and budget.

The glistening squid put on a dazzling technicolour show, using their chromatophores to send out pulsating pixels of light. They were carefully transferred in colourful plastic pails and passed on to some very experienced, leathery hands. We availed ourselves of our status as privileged customers and interested onlookers to form a friendly relationship with the women handling the cephalopods, which allowed us to learn the many secrets of how the squid are cleaned, prepared, and spiced.

The squid, which varied in size from five to fifteen centimetres, were stored in fresh seawater. They were scooped up with a plastic spoon or tongs and the ink sacs and pens were extracted in a flash. Then they were put on bamboo skewers inserted through the mouth and woven through the mantle. The small octopuses are not usually very sought after because their meat is a bit tough and hard. Nevertheless, they too were cleaned and the innards removed before they were placed on a skewer that was inserted at the top of the mantle and came out through the mouth. Because the meat is thicker they were grilled in two stages. First they were placed on the grill whole and turned over. Then they were refastened on the skewers very decoratively so that the arms hung down over the mantle, a bit like someone in a dress doing a cartwheel, and grilled again. With input from the onlooking guests they were marinated right on the charcoal grill with a pre-mixed marinade made with oil, chili, crushed Kampot pepper, salt, lime leaves, lemon grass, fish sauce, garlic, and a bit of MSG, or a selection of these ingredients according to taste.

The warm squid were placed on a take away banana leaf together with a small bag of sauce made from lime juice, fish sauce, sugar, and chili. We carried this package a bit further into the market where we could buy extra sauces or vines of fresh green Kampot pepper, which a local English-speaking Cambodian had recommended that we should crush onto the squid.

All that was left was to buy a cold beer from the table owners as payment for the use of their plastic covered tables and some much needed serviettes. It was the start of a meal where every single mouthful was like eating the sea itself.

One becomes especially aware that pepper is not just any old pepper when one has tasted and smelled the freshly harvested green pepper from a place like the Kampot region of southern Cambodia. We were able to visit Sothy's Pepper Farm, which is located in a fertile area that is surrounded by an enormous nature reserve.

The tradition of pepper farming is due to the influence of China and goes back to the 1300s when it was discovered that the special type of mineral-rich soil in the area is ideally suited to cultivating these plants. This pepper is now regarded as among the best in the world.

The combination of grilled cephalopods and fresh green Kampot pepper is wonderfully synergistic, in fact, pepper sauce was about the only condiment for these dishes that we encountered in Cambodia. It is easy to prepare. Lime juice is simply mixed either with one tablespoonful of green peppercorns crushed in a mortar or with ground, not too old, black Kampot peppercorns to make a fairly thin dressing.

Our first experience of the restaurant Chili Pepper Khmer Cuisine was the start of many return visits. It was one of those serendipitous occasions when a sudden jolt of the right chemistry between the drawing power of the restaurant and our search for the best local gastronomic experiences meshed perfectly. We were fortunate to engage in a good exchange with the head chef, Dorn Doeut, who introduced us to the new style of Cambodian cuisine, which respects its origins and traditions but is just as exciting as any of the novel approaches that are making inroads in the rest of the world.

◘ Local fishers on Koh Rong Island in Cambodia sorting the nightly catch early in the morning.

Cambodian cooks are blessed with access to a wealth of raw ingredients such as crabs, fresh and salt water fish, cephalopods, shellfish, many kinds of ripe fruits, lemon grass, coconut, both green and ripe mangos and jack-fruit, a huge variety of wild herbs, and banana flowers. To that one can add many taste enhancers—lime leaves, fish sauce, fermented fish, dried fish and shrimps, and Kampot pepper. We were impressed by the abundance and high quality of the ingredients and the way in which they are seasoned. The result is a delicate, elegant of cuisine in which the various tastes can be clearly distinguished from each other. This is very different from the much spicier approach in the culinary traditions of nearby countries.

The Cambodian national dish, *amok*, variations of which are found as *mok* in Laos and *ho mok* in Thailand, is an exceptionally delicious dish, which dates back to the royal Cambodian Angkor-Khmer kitchen. The recipe is generally passed on from one generation to another within the family and is subject to fierce debate with others about the correct ingredients and how it should be prepared. Normally it involves steaming local fresh water fish in banana leaves. Its consistency varies from soft to a little full in a creamy sort of way and it has a wonderful aroma. Eating *amok* is a unique taste and sensory experience that is due to the special *khmer-kroeung* spice mixture. Each delicious mouthful seems to be the embodiment of a thousand years of Cambodian culture, history, and knowledge.

The recipe below is an example of Cambodian *amok* made with Japanese flying squid—it is so full of wonderful Asian flavours that it literally gives one goose bumps. The elegant, meaty mouthfeel of the squid creates a fantastic contrast with the creamy *amok* sauce.

Amok made with squid

Serves 4

600 g (1 1/3 lb) Japanese flying squid (*Todarodes pacificus*)

1 Tbsp coconut oil or other neutral tasting oil

125 g (1/2 c) *kroeung* (recipe below)

4 dL (1 2/3 c) chicken broth

4 dL (1 2/3 c) fresh or canned coconut milk

2–3 Tbsp fish sauce

1 Tbsp palm sugar

1/2 dL (2/3 oz) *noni* leaves

To serve

Fresh red chili pepper

Fresh lime leaves

1 dL (4/5 c) coconut cream

Fresh banana leaves

1. Cut the mantle of the squid into 16 square pieces and score them in a diamond pattern and set aside.
2. Put the coconut oil in a saucepan, heat it, add the *kroeung* mixture, and toast lightly.
3. Add the chicken broth, coconut milk, fish sauce, sugar, and *noni* leaves. Allow to simmer for 20–25 minutes on low heat.
4. Add the squid pieces and simmer for a further 5–8 minutes until they are just done.
5. Season, adding the toasted *kroeung*, fish sauce, and possibly a little salt as needed.
6. Cut the fresh chili and lime leaves into thin strips. Just before serving whip the coconut cream to a foam while warming.

7. Arrange the piping hot *amok* in an attractive bowl. Optionally the serving dish can be decorated with banana leaves that have been softened in water and cut to size.

8. Distribute the coconut cream foam evenly on top. Sprinkle with the chili and lime leaf strips.

Cooked rice is an excellent accompaniment for the dish as one wants to be able to soak up all the liquid in which the squid was simmered. *Noni* leaves (*bai-yo*) may be hard to get outside Asia. Fresh lime leaves can be used instead.

◻ The Cambodian national specialty *amok*, in this case prepared with squid.

Kroeung spice mixture for khmer-*amok*

350 g (3/4 lb) fresh lemon grass
50 g (1 3/4 oz) fresh turmeric
100 g (3 1/2 oz) fresh ginger root (or fresh galangal)
75 g (2 2/3 oz) garlic
50 g (1 3/4 oz) shallots
50 g (1 3/4 oz) dried chili, softened in warm water
10 g (1/3 oz) lime leaves
0.1 dL (2 tsp) fish sauce
0.1 dL (2 tsp) palm sugar

1. Remove the outer leaves of the lemon grass and cut up finely.
2. Peel the turmeric and ginger and grate them both.
3. Peel the garlic and slice it finely.
4. Peel the shallots and dice it.
5. Slice the lime leaves into thin strips.
6. Crush all the ingredients, including the chili, in a mortar or chop in a food processor. Add the fish sauce and palm sugar. Store the mixture in a cool place until needed.

   If one uses a food processor, the ingredients do not have to be chopped as finely to start.

This recipe makes a fairly large portion, as it is almost impossible to make *kroeung* in an amount that is just enough for a single dish. But this is not a problem as it can be stored in a cool place or in a freezer for two to three months.

Kroeung is quite versatile and can also be used with fish, fowl, pork, soup, and vegetable dishes, as well as in a dip.

Gastrophysics and 'The Squid Squad'

The long-finned squid (*Loligo forbesii*) and common squid (*Loligo vulgaris*) are commercially important species found in abundance in all the seas around Europe, including the northern waters bordering Scandinavia. But as they have never really found a place in traditional Danish cuisine, there was a tangible incentive to study this food source in depth in an attempt to raise its gastronomic profile in Denmark.

To this end a group of researchers, students, communicators, and chefs, who came to be known as 'The Squid Squad,' got together in the fall of 2017 at the University of Copenhagen to carry out, both in the laboratory and the kitchen, a series of scientific studies based on gastrophysics. The underlying idea was not simply to research different ways of preparing these two species of squid in such a way that they would have an interesting taste and texture that would appeal to Danish palates. It was also to draw attention to this sustainable, yet underexploited, low-fat protein source that could be adopted as an alternative to meat from terrestrial animals. Both aspects of this work have great potential to stimulate innovation and lead to the development of new products. In parallel with this effort, the work of the group was linked to outreach to school children, young people, and the general public under the aegis of the Danish research and communications centre, *Smag for Livet* (*Taste for Life*).

What Is Gastrophysics?

Gastrophysics is a relatively recent, interdisciplinary subject that is empirically based on the exploration of food cultures, raw ingredients, and the culinary techniques that maximize their taste and nutritional value and on how these can be studied in a rigorously scientific way based on the principles of physics and chemistry. In concrete terms, a gastrophysical approach starts with a specific gastronomic observation or a curiosity-driven question that arises in relation to the actual preparation of food in the kitchen, followed by quantitative, experimental, and theoretical analyses of the process in a gastro-lab. These, in turn, can result in a new insight, for example, into why a food has a particular taste and texture. Armed with this knowledge one can then return to the kitchen with suggestions for improvement, a new recipe, or, possibly, the introduction of a new product.

'The Squid Squad'

Given how well squid checked the boxes of availability, sustainability, and underrepresentation in Danish cuisine, they were a natural fit for a multi-faceted study. This work was undertaken in the fall of 2017 by a group made up of researchers, food science engineers, students, and chefs, who were nicknamed 'The Squid Squad.' Their goal was to use gastrophysics to explore how to maximize the potential of locally caught *Loligo forbesii* and *Loligo vulgaris* as a food source. The project is described from their perspective in the text below.

With regard to squid, the obvious gastronomically inspired question that formed the basis of the study was: How can squid be prepared to achieve the desired texture and taste? In addition, it was possible to combine the gastrophysical investigations with sensory studies. These could give us an insight, on the one hand, into the interaction between how squid are prepared (and related measurements of material properties, taste, and aroma substances) and, on the other hand, how a panel of sensory experts would describe the taste experience and their eventual preferences.

We have inserted three recipes that were developed by members of 'The Squid Squad' as concrete examples of the novel ways devised by the chefs to prepare the cephalopods. Admittedly, two of them require ingredients that it might be hard for the average home cook to find and also call for sophisticated professional equipment. The rationale for including them was to demonstrate the knowledge gained about the extent to which squid ingredients can be used creatively, some of which might ultimately be incorporated into ready-made products for domestic consumption.

'The Squid Squad.' The members of the group were Louise Beck Brønnum, Charlotte Vinther Schmidt, Yi-Ting Sun, Peter Lionet Faxholm, Roberto Flore, Karsten Olsen, and Ole G. Mouritsen. The description of their project in what follows is based on a previously published, and more detailed scientific paper.

Putting Nordic Squid to the Test

The high season for squid in the waters around Denmark is November to January and that was when we took delivery of excellent, very fresh Nordic squid, especially *Loligo forbesii* that had been caught in the North Sea. At the beginning of the season their average weight was about one kilogram and toward January it was about one-half of that. The largest specimen we examined was a 4 kilogram male, with a total length of about 135 centimetres and a

Because squid are cannibals, they tend to hunt in schools made up of individuals of a similar size, which minimizes the chance of large ones preying on the small ones. As a result, when a school of squid is caught up in a fisher's net there is often a bias toward squid of the same sex.

There is little equality between the sexes among cephalopods. In the case of octopuses, there are typically three times as many females as males and, for some species, the ratio can be up to 50 to 1. The reverse is true for squid where there are three times as many males as females. For cuttlefish, there is approximate gender parity.

mantle measuring 63 centimetres. The deliveries from the supplier tended to contain more males than females, with all the larger ones being male.

The squid were cleaned promptly after delivery and dissected into their various anatomical parts—mantle, arms, tentacles, fins, head with the eyes, buccal mass with the beak and tongue, liver (hepatopancreas), retractor muscles, siphon, and ink sac. In the case of females, the reproductive organs consisting of nidamental glands and eggs were also collected. The digestive system, gills, and hearts of both sexes and the male reproductive organs were regarded as offal and discarded. In some cases, the middle part with the eyes was separated out and dried. The cleaned pieces were packed in plastic pouches and frozen for later use. Samples destined for microscopy and chemical analysis were stored at −40° Celsius. Those that were used in gastronomical experiments and evaluated sensorially were kept at −20° Celsius or lower for twenty-four hours.

The studies of squid gastronomic potential proceeded along two parallel tracks and the results were compared as they unfolded. One track was primarily under the direction of chefs, who initiated a variety of approaches to preparing the squid. The other was overseen by scientists, who tested samples of the various parts of the squid with regard to a physical analysis of their texture, a chemical analysis of their nutritional components, and a later microscopic examination of the collagen structure of the muscles.

ⓘ Silky, curly squid strips

Serves 4

1 squid mantle (*Loligo forbesii*),
   weighing ca. 800 g (1 3/4 lb)
1/2 L (16 fl oz) safflower oil
1 clove garlic, crushed
Fresh parsley, chopped
Salt and pepper

1. Using a sharp knife, slice the mantle into lengthwise strips approximately 1–2 mm (about 1/16 inch) across. The thinner the strips, the more they curl up in the hot oil. It is easier to slice really thin strips if the mantle has been lightly frozen beforehand. If preparing this recipe in larger quantities for a big group, it might be an idea to use a meat slicer.

Several mantles can be layered on top of each other, frozen lightly, and then all sliced at the same time.

2. Place the strips in a pot with the crushed garlic and cover with the safflower oil. The pot should be big enough to allow the strips to move around and curl up when they are heated.

3. Heat the pot quickly over medium high heat until the oil reaches a temperature of 85 °C (185 °F). As it heats, stir the strips occasionally so that they have a chance to curl up. This should take a maximum of 5–10 minutes. When all the strips have curled up, keep the temperature of the oil at a constant temperature of 85 °C (185 °F) until the strips are silky soft and tender. This should take about 35–40 minutes, depending on the size and thickness of the mantle. In the case of very large squid it might take an extra 15 minutes before the pieces are tender.

4. Remove the curled up strips from the oil with a slotted spoon and place them on paper towels for a short time to absorb the extra oil.

5. Place on serving plates, sprinkle the chopped parsley on top, and season to taste with salt and pepper.

To serve

This dish can be served as quickly as possible while the squid is still hot. The strips can also be served after they have cooled, or they can first be tossed in a sauce or a dressing. The cooked garlic also tastes quite wonderful.

Variations

Instead of safflower oil, one can use other vegetable oils (for example, olive, rapeseed, or canola oils) or clarified butter. The parsley and garlic can be replaced by other combinations such as rosemary, seaweed, nuts, and seeds. To make an Asian-inspired version, add coriander, star anise, a cinnamon stick, bay leaf, and cloves to the oil. Toss the strips in a soy/honey glaze and sprinkle with fresh chopped coriander and toasted sesame seeds.

◘ Silky, curly squid strips served with chopped parsley and a lemon wedge. Because of the difference in the musculature at the surface and of the inner part of the mantle, the finely sliced strips contract unevenly causing them to curl up as they are warmed and they end up with a silky, soft texture. (Recipe: Peter Lionet Faxholm).

The gastrophysical analysis of the texture of the various parts of the squid was carried out by mechanical techniques to measure the force needed to deform the squid meat and identify the point at which the meat starts to break up. This can be compared with the effect of the force that is applied by chewing movements. The results of such measurements are described using terms such as hardness, elasticity, cohesion, springiness, and toughness. This allowed us to obtain precise information about how a particular treatment affects the tenderness of arms and mantle. Determination of the chemical compounds was carried out using analytical techniques that are able to identify them and measure the quantities of each. We also focused on singling out the aroma substances that give rise to the characteristic, but possibly unpleasant, fish smell and the taste substances that are responsible for umami.

As the work was carried out in a gastro-lab, it was done in strictly systematic fashion, accompanied by careful documentation, and consistent attempts to achieve results that were reproducible. At the outset, there were numerous conditions that could be varied: which part of the squid was being studied, the temperature, the time taken to prepare and store the sample, water content, the effect of salt, acid, enzymes, smoking, and fermentation processes, and so on. Obviously, it is theoretically possible to combine these various conditions in so many ways that it would be impossible to test all of them. This is where culinary intuition and scientific expertise came into the picture. The challenge was to select those combinations that were expected to be most relevant and interesting from a gastronomic perspective, without shutting the door on the unexpected and spontaneous results that could follow from an experimental approach.

ℹ Black sauce made with squid ink and squid livers

5 ink sacs or 20 g (4 tsp) squid ink
200 g (7 oz) squid (*Loligo forbesii*) liver
5 g (1/6 oz) chipotle pepper
3 g (2/3 tsp) coriander seeds
70 g (2 1/2 oz) apples
150 g (5 1/4 oz) onions, peeled
10 g (1/3 oz) garlic, peeled
120 g (5/8 c) dry white wine
3 L (100 fl oz) *dashi*
2 bay leaves
Salt and olive oil according to taste

1. If using squid ink sacs, carefully extract the ink and set aside.
2. Toast the liver, chipotle, and coriander seeds with salt and oil on a warm skillet for about 10 minutes.
3. Peel and chop the apples into small pieces. Add them, together with the onion, bay leaves, and garlic to the liver mixture. Add half of the ink and allow to toast for a further 5 minutes.
4. Add the white wine. When the alcohol has evaporated, add two-thirds of the *dashi* and allow all to simmer until the mixture is almost dry.
5. When the mixture has cooled put it in a thermomixer at top speed for 4 minutes.
6. Using a spoon, press the mixture through a sieve. Put the resulting purée in a pot with the remaining *dashi* and the rest of the ink. Cook gently until it has reduced to a thick sauce.

This sauce is versatile and can be used many ways, for example, as a dip for snacks or as a sauce with fish and cephalopod dishes.

🔲 Black sauce made with squid ink and squid livers. (Recipe: Roberto Flore).

🔲 The two parts of the beak of a *Loligo forbesii.*

The different ways of preparing the squid included the following: freezing, sous vide heat treatment, partial drying and full dehydration, marinating in salt, acid, or squid ink, fermentation using the squid's own intestinal enzymes (*shiokara*), fermentation with *koji*, smoking, baking, sautéing in oil, grilling, and grinding to make a thick paste.

Along the way we gained an understanding of the optimal temperatures for heat treatment, the applicable lengths of time for fermentation and marinating processes, and, not least of all, which of the different techniques for tenderizing the various parts of the squid meat were most successful. Much to our surprise, we discovered that, contrary to the claims of many chefs, freezing made no noteworthy contribution to changing the texture of the squid, at least to the extent that it was possible to characterize quantitively using physical techniques. We did observe a slight loss of liquid from the frozen samples. Nevertheless, this loss was either of less importance in terms of texture or the freezing might have had a compensating tenderizing effect. All of this is in contrast to what happens with octopuses, as freezing makes their meat noticeably more tender.

Among our more practical goals were attempts to come up with new products using cephalopods that could appeal to the Danish palate, for example, as dried snacks or in cans. On the more demanding culinary end of things we were striving to develop new dishes with the tastes and textures that enhanced the particular characteristics of these raw ingredients.

◘ Prototype of a new canned product developed by 'The Squid Squad.' The can contains prepared squid arms from *Loligo forbesii* in oil. The label is designed on the basis of a *gyotaku* imprint of the tentacles of the squid. *Gyotaku* is an old Japanese art form by which one makes an image of a fish or cephalopod by painting the surface of the animal with *Sepia* ink and the pressing a piece of absorbing white paper on top of it.

Some Surprising Results

When chefs and scientists get together to conduct collaborative experiments they sometimes generate wild ideas and inventive new ways to prepare the ingredients that are very different from anything that has ever been seen. An example of this is Roberto Flore's design of a braid of steamed, lightly smoked, and cured tentacles. After he had described this idea and illustrated it with a picture on social media, other chefs in Denmark and around the world ran with the concept. Just as cephalopods can jet-propel themselves in the water, it appears that new and inspiring ideas about how to prepare them travel even more quickly in cyberspace.

■ Roberto Flore's design for a braid made from steamed, lightly smoked, and cured tentacles of *Loligo forbesii.*

Another example is based on a sensory surprise. When one is working with the natural texture of the squid it is reasonable to imagine an extreme case of virtually eliminating it. For example, the meat can be blended or put through a Pacojet until it is turned into a paste. It turns out that the resulting paste is smooth and creamy and that the natural sweetness of the squid meat comes forward.

This raw squid paste inspired Roberto Flore to turn to his culinary roots. In Sardinia there is a type of traditional toasted flatbread, *pane carasau,* that the local shepherds used to take along when they were away from home for months at a time tending their sheep. *Pane carasau* is made from durum wheat and baked twice. The dough is rolled out into very thin sheets that are first baked and then brushed with olive oil, placed one on top of the other, and then baked again until they are very crisp. Normally, these sheets are very large, up to one metre wide and pieces are broken off as needed. In Robert Flore's version, the *pane carasau* were small round shapes, much like a normal sweet biscuit.

❶ Sardinian-inspired squid biscuits

Makes 10 pieces

100 g (3 1/2 oz) squid mantle from *Loligo forbesii*, cleaned

40 g (1 1/2 oz) Sardinian *pane carasau*

15 g (1 Tbsp) butter, melted

10 g (2 tsp) powdered sugar

100 g (3 1/2 oz) dried lemon verbena

100 g (2/5 c) cold pressed rapeseed oil

5 g (1 tsp) salted capers

1. Be sure that the skin is completely removed from the mantle. Quick freeze it and cut it up into smaller pieces. Place the pieces in a Pacojet, let it run twice, and keep the paste in the freezer until ready for use.

2. Moisten the *pane carasau* and use a cookie cutter to make shapes that are about 4 cm (1 1/2 in) in diameter. Brush the biscuits with the melted butter, sprinkle powdered sugar over them, and glaze them in the oven at 200 °C (395 °F) for 4 minutes. Allow the biscuits to cool.

3. In a Thermomixer blend the dried lemon verbena and the rapeseed oil at 70 °C (160 °F) for 7 minutes and then allow it to rest for 6 hours. Sieve the oil, place it in a vacuum-sealed bag in a refrigerator, where it can be kept for later use. For this recipe only 5 mL (1 tsp) of the oil is used, but it is impractical to make it in such a small portion.

4. Rinse off the excess salt from the capers and dry them in a dehydrator or an oven at 50 °C (120 °F). Blend them to a powder.

5. Top the *pane carasau* biscuits with a scoop of the frozen squid paste. Drizzle with a few drops of the lemon verbena oil, dust with the caper powder, and serve immediately.

A glass of good Sardinian Malmsey pairs perfectly with this dessert.

◨ A Sardinian inspired dessert biscuit consisting of a piece of *pane carasau*, the traditional Sardinian twice-baked flatbread, topped with a frozen paste made from the blended mantle of raw *Loligo forbesii*.

Cephalopods Come into Their Own at *Taste for Life*

'The Squid Squad' carried out its project on local cephalopods as a collaborative effort with the national Danish communication and research centre *Taste for Life*. The work of this centre is largely focused on disseminating knowledge about the taste of food to school children, young people, and the general public. One of the approaches taken at the centre involves offering the opportunity to sample strange or exotic foods with unusual tastes and textures, for example, innards, jellyfish, insects, and seaweeds. The aim is to encourage individuals to become more curious and willing to experiment, to break out of their comfort zones and try new things, and to develop their own unique notions of what is palatable.

So it was only natural that Nordic cephalopods should become the subject of some of this outreach. In the beginning of 2018 arrangements were made to hold different sessions that focused on their taste. One was for curious foodies, another for a class of middle school children, and lastly one for a group of fishers, fish vendors, and chefs.

The session for curious foodies took the form of a kitchen talk between a chef and a scientist with expertise in ocean-based ingredients, fermentation, and food safety, who were joined by a bartender who could provide samples of wine pairings. The audience was able to learn more about the cephalopods, see them close up, and taste them in the dishes prepared by Roberto Flore. The underlying idea was to demonstrate that the concept of palatability could easily be expanded to include squid caught in local Nordic waters.

The session geared to school children was attended by students from grades seven and nine who had chosen an elective course in Food Science. They took part in an activity called 'Know Your Squid' where they actually handled the squid and made food with them. In accordance with the syllabus for this course, this activity was designed to allow the students to experience first-hand, and understand the significance of, the physiology of a particular animal and how it is transformed to be viewed as a food source. The pupils' reactions to this experience were subsequently reflected in their overall taste impressions of the squid and a greater willingness to sample and eat squid. The combination of a biological approach to an ingredient in its animal state and the gastronomic utilization of the animal in the kitchen provided a foundation for further work on the taste and texture of cephalopods.

The workshop for fishers, fish vendors, and chefs was intended to focus on the current situation that there is really no market for Nordic squid within the country, even though there is a great abundance of them in the surrounding waters, especially in the North Sea.

The work undertaken to put together this book and the efforts of 'The Squid Squad' to expand gastronomic horizons have demonstrated that it is very possible to make food that tastes good using this resource. It is to be hoped that these endeavours and the subsequent outreach activities will bring about a domestic demand for Nordic squid. But in order to create a seamless link from ocean to table it is necessary for the fishers, fish vendors, and chefs to become aware of the possibilities and each do their part to promote cephalopods within Danish food culture. That was the true aim of the workshop.

◘ Cutting up a large Nordic *Loligo forbesii* in front of a group of school children.

A Field Trip to Experience Cephalopod Fisheries and Cuisine in Sardinia

Thanks to the arrangements made by a member of 'The Squid Squad,' Roberto Flore, two other members of the group (Louise and Ole) had a great opportunity to visit Sardinia to experience the many ways in which cephalopods are integrated into the island's food culture.

Roberto Flore was head of culinary research and development at Nordic Food Lab from 2014 when he arrived in Denmark until his move to the Technical University of Denmark in 2019. Apart from being a trained chef, he has completed agricultural studies and he takes a particular interest in using marine food sources in a sustainable manner. Roberto was born in Sardinia, where he became fascinated with food and gastronomy as a small child. His earliest memories are of spending time in the kitchen with his grandmother while she kneaded bread, made ravioli, and baked sweet cakes. He would go along with her into the surrounding landscape to forage for plants and herbs. He grew up in the Sardinian food culture which values the raising of pigs, hunting and fishing, and the olive harvest. Roberto still makes his own wine and olive oil back home.

Our field trips covered most of the important aspects of cephalopod cuisine in Sardinia. We first encountered them as raw ingredients at the seafood market and Roberto had also arranged for us to go to sea with the local fishers to catch the creatures that we would later be eating. We were able to meet and exchange ideas with Sardinian chefs who prepared both simple traditional and avant-garde dishes for us. It turned out that we had something to teach them in return.

We arrived at the entrance to Sardinia's largest fish market in the capital city of Cagliari at half past three on a Tuesday morning in the beginning of February, to be there as soon as it opened. Fishers come at night from all over Sardinia to sell their catch at auction to the wholesalers, who in turn supply the retail fish stores and restaurants. Naturally, they are bringing not just fish but also cephalopods and shellfish, including the highly prized Mediterranean red shrimp. In the boxes in the large hall we could see octopuses, different varieties of cuttlefish, for example, *Sepia officinalis*, and squid, especially *Loligo forbesii* and *Loligo vulgaris*. We were accompanied by Mario Mangano, a fisher who would take us out on his boat that afternoon, and Guiseppe Ollano, the foreman of Peschiera della laguna di Nora, which we were to visit the next day. Apart from the fishers and all the fresh seafood, the market was jam-packed with chefs and retailers who were practically racing around the hall to get their hands on the very best of what was on offer.

Pier Luigi Fais, chef-owner of a fine-dining restaurant, Josto, who was going to prepare a special cephalopod dish for us the next day, had also arrived and the conversation turned to the cost of the fish. Locally caught seafood makes up only about 30 percent of what is sold at the market, with the rest coming from other countries. The dynamics of the market are strongly affected by the prices of the imported products, which are roughly half of what is charged for the local ones. Everyone was in agreement, nevertheless, that the Sardinian calamari are better than the imported ones. Roberto said that he preferred cuttlefish, but Pier Luigi would rather eat squid (*Loligo* spp.) because they are the sweetest and he also uses their liver to make special dishes.

It was still very early in the morning so we returned to the hotel to nap for an hour or so before heading out to have some breakfast, which turned out to be a very unusual sort of breakfast. We went to Mercato di San Benedetto, one of the biggest fresh produce markets in Europe, where the retail sales of seafood take place. The market, which is divided into three levels, was a hive of activity as we made our way to the booths of the fish vendors in the basement. If one is a fan of all good things from the sea, this is a veritable paradise—the goods looked fantastically fresh, having been caught at night and brought from the wholesale market just a few hours previously. A couple of the vendors recognized us as the strange people who had been poking around the fish market earlier that morning looking in all the boxes but who definitely did not belong there.

We were naturally drawn to the huge selection of cephalopods and were suddenly seized by the idea of buying, on the spot, a large squid (*Loligo vulgaris*) and eating it raw for breakfast. One of the vendors went along with the plan and directed us to a small sushi booth next to his own, where a lady sushi chef might be persuaded to cut up the squid's mantle to make sushi and *sashimi*. She agreed and after the vendor had shown us how he cleaned and divided up the squid, we took it all over to the sushi chef. She quickly cut it up so that, after being up since three o'clock in the morning, we could finally have our well-deserved breakfast: *ika-sushi* and *ika-sashimi*. It could not possibly have been better!

Late in the afternoon we drove about forty kilometres southwest from Cagliari to the coast at Nora. Here we were meeting up with Mario and his colleague Amerigo Sandolo, who were going to take us along to put out nets to catch fish for our lunch the next day. This area has been inhabited since prehistoric times by people from other parts of Europe who settled there and later by the Phoenicians who had conquered coastal Sardinia until they, in turn, were defeated by the Romans in the third century BCE. In Roman times Nora became an important port on account of its location on a promontory that provided a safe harbour on three sides. Most of the ruins of this ancient city have now disappeared under water as the littoral continues to sink in this part of the island.

When we arrived Mario and two of the other local fishers were busy cleaning the nets. Because the water close to shore is very shallow, the fishing boat was lying at anchor about fifty metres out. So we were put in two small, beaten-up rowboats and, with the help of oars and an old punt-pole, ferried out to it. Once we were on board, where we could barely stand up on the deck of the small, blue-painted boat because it was covered with large piles of gear, Mario set the course to head out to where the nets were to be lowered. He had cautioned us not to get our hopes up that there would be a good catch as a storm the previous day had stirred up the water and it was not very clear.

On the way we stopped by a buoy where a couple of plastic bottles were bobbing up and down on the waves. They were attached to their *nessa* that were lying on the seabed. A *nessa* is a special cylindrical octopus trap made out of netting. The trap has a narrow opening and once an octopus has entered it cannot get out again. The bait is a crab, which is the favourite food of octopuses. The traps are checked once a week and we were eager to see whether there was anything in them. Amerigo hauled up the line and a series of traps slowly came up from the depths. The first one was empty, but the next one contained an octopus which was dumped out on the deck. It was a white-spotted octopus (*Callistoctopus macropus*), which has very long thin arms. It immediately put its sensitive arms and their suckers to work to explore its strange new surroundings. We put it into a bucket, but we had to keep an eye on it to prevent the octopus from escaping.

Other *nessa* had trapped some common octopuses, *Octopus vulgaris*. They were crawling around on the deck when Mario suddenly grabbed one and bit it behind its eyes. This rendered the octopus brain dead, in the sense that the central brain had been decoupled from the rest of the organism. In a flash the animal changed colour and turned ash grey, because the brain had lost control over the many small muscles that surround the chromatophores and cause it to change colour. As the muscles relax, the small colour sacs contract, and the red colour disappears. But the arms continued to move somewhat even though the octopus has been immobilized. Mario explained that biting the octopus was in some senses humane because it did the least damage to the animal.

Mario then described how he manages on the days when he is alone, as it can be quite dangerous to fish on one's own in the Mediterranean. He stands at the back of the boat with the rudder between his legs while he hauls in a net. If there are octopuses in the net, it is necessary to immobilize them immediately so that they do not crawl away. It is much safer for him to bite them quickly than it is to bend down and use a knife to kill them, especially if there is a swell. A few years ago Mario lost a brother, also a fisher, who was a year older than him. He fell overboard and his boat drifted in to shore without him. His body was never found.

In the late afternoon Amerigo and Mario started to put out the nets in water that was about ten metres deep. After an hour or so, the work was done and the sun set, bathing the old port at Nora in a beautiful reddish light, as we headed back to shore. We were in a bit of a hurry as we needed to deliver the freshly caught octopuses to Luigi Pomata, who was going to prepare a traditional meal for us that evening back in Cagliari.

Luigi, who is a good friend of Roberto's, is a well-known Sardinian TV chef and owner of a highly rated restaurant that bears his name. He presented us with a true Sardinian seafood feast that included both fish and octopuses. Believe it or not, we started with canned tuna in good olive oil, served in its square blue tin placed right on the table. It was not, however, just any old can of tuna. It contained the finest belly meat (*ventresca*) from bluefin tuna caught near the island of San Pietro, just off the south-west coast of Sardinia.

Among the other items on the menu, there were naturally also ones made with octopus. We were served two very different, traditional ones. The first was octopus arms that had been pressure cooked, served in a cephalopod ink-balsamic reduction and with *farinata*, a baked chickpea mash. The arms on the plate were from 'our own' octopus that had been caught on Mario's boat that morning. Luigi pointed out, though, that ideally the meat should have been frozen overnight to ensure that it was as tender as possible. The second one consisted of black *gnocchi* with squid ink and sea urchins, served with *burrata* and a mandarin sauce.

The next morning there was no time for breakfast either, as we had to leave for Nora very early to tend the nets. The wind had turned and where there was calm water the previous evening, it was blowing hard and there were threatening, dark clouds overhead. But that did not put us off. Mario said that it would rain and gave us some waterproof red oilskins for protection. On the way out, the waves were high, forcing us to keep a tight grip on the hawsers and fittings, as the gunwales are very low. We reached the plastic bottles that marked the location of the nets and were eager to find out whether they were full or empty.

There were a few fish in the nets before we got our first cephalopod, a small cuttlefish, *Sepia officinalis*, which shot out some of its ink as it landed on the deck. Then came a few more fish, a pair of small octopuses, more cuttlefish, some of which were quite large, and eventually the first squid, a *Loligo vulgaris*. This was the moment we had been waiting for. It was a large, impressive specimen and Amerigo carefully disentangled it so that we could see and touch it. It was almost as transparent as glass and its skin, with the many small, red spots with chromatophores, was perfect and undamaged by the net. It was an incredibly beautiful creature and one can only imagine how elegant it would look in its natural element in the water.

We were not able to see the squid's tentacles as it had retracted them right back toward its mouth. But Amerigo grabbed hold of one of them, stretched it out to its full length, and then bit off a piece in the middle, where there are no suckers. Apparently, this is something that fishers do now and again as, much to our surprise, they like the taste of raw *Loligo*. Naturally, we also had to try some and the poor animal was passed around so that we could all taste the tentacle. The meat was soft, elastic, and easy to chew. It tasted sweet and creamy, with a pleasant aftertaste of the fresh seawater.

One of us (Ole) also wanted to try to eat the club, that is to say, the end of the tentacle with a lot of suckers. This means having to chew really fast to prevent the suckers from attaching themselves to the membranes inside one's mouth. The chitin rings on the rim of the suckers create an interesting crunchy feeling between the teeth. On the one hand sitting on a boat and eating part of a still-living creature seemed a little like crossing into unexplored territory, but on the other hand it felt quite natural. The squid was placed in a pail that was separate from the one with the octopuses. The fishers maintain that being next to squid gives the octopus an off taste.

The fishers' pessimism about the size of the catch proved to be accurate as many fewer fish and cephalopods were caught than on a good day. But little by little, as the nets came in, we saw all three types of cephalopods in the mix. As that was what we, the amateur fishers, had hoped for we were feeling very satisfied. Suddenly a large *Sepia officinalis* was pulled in and as Amerigo freed it from the net, it squirted a stream of black ink right at the heads of Louise and Ole. This truly fulfilled our aim of getting an up-close and personal experience of these creatures and it proved to be the highlight of our trip.

The weather cleared up and the sun came out as we set course for our return to port. Once there, we unloaded the boxes with the catch, which we were taking on to their next destination, Peschiera della laguna di Nora. It is housed in some buildings on a narrow strip of land that separates the lagoon from the sea. We were met by the managers of the *peschiera*, Giuseppe Ollano and his wife, Daniella Fadda, together with the chef Davide Atzori. Davide, who at the age of thirty-three, had recently been put in charge of the *peschiera*'s restaurant, Is Fradis Minoris, which at the time was closed for the season. But the kitchen had been opened up especially for us so that he could prepare a lunch with the cephalopods that we had just caught.

Davide had previously worked in highly regarded establishments in Paris, but had decided to return home to Sardinia to fulfil his dream of running a kitchen on the basis of seafood that was harvested locally and sustainably with respect for the environment. What he missed in Paris was direct contact with the producers who deliver the raw ingredients to the restaurants. Not so in Nora where, as he told us, "Here I make food from what Mario catches."

These underlying ideas about bringing together the fishing industry and sustainable development with a nod to the area's history were also what motivated Guiseppe and Daniella, who are both biologists, to come up with plans some 30 years ago for a project that eventually resulted in setting up the present day Pescheria di laguna di Nora. Guiseppe explained that a *pescheria* was originally a type of co-operative that looked after the land, the coast, and the community that supported itself from its fishing rights in that area. In the case of Nora that included the lagoon, which is cut off from the sea by a narrow, low-lying strip of land. From ancient times until recently fishers took advantage of this configuration to catch marine fish, such as sea bass, in the lagoon. After the fish entered the lagoon at high tide, the fishers lowered grills into the opening to trap them when the water rushed out again at low tide.

Their proposal faced a major stumbling block by way of opposition from the surrounding community that wanted to build a marina for tourists. Guiseppe did not hide the fact that it had been an uphill battle to bring the local government and people on board and gain an understanding of the new, modern concept. But after many years and with the support of the Italian government, among others, public opinion was slowly turned around. The resulting many-faceted organization now involves the fishers themselves, provides protection for the natural environment, runs a museum, an interpretive centre, and a restaurant, and supports the tourist industry. The project is regarded as a shining example of the commune of Pula's strategy to develop tourism as a source of employment.

A vital part of the project is the restaurant, which exclusively serves local seafood, such as sea bass, grey mullet, and common cockles that are caught in the lagoon and off-shore. There is an emphasis on using only the fish and shellfish that can be harvested in accordance with certain regulations and in a sustainable manner and on selling all of the catch to the restaurant. One aspect of the restaurant's mission is to encourage its patrons to eat species that are not part of the traditional food culture and to minimize food waste. Even though the diners request them, the menu does not feature the very popular, but endangered, scampi. This decision reinforces another facet of the project, which is to foster an understanding in the visitors about the philosophy that underpins establishing a sustainable ecosystem. Mario is responsible for the fishing operations and it is first and foremost his catch that supplies the restaurant.

The challenge in managing a restaurant in this way is that it is not until about nine or ten in the morning that the chef learns what Mario has caught. Only at that point can the menu of the day be set, actually much like the day when we were the guests and brought our own freshly caught, raw ingredients with us. While we stood around in the little kitchen Davide talked about his approach to preparing seafood. He said that in his world there are neither good nor bad fish. Rather it is mostly a matter of an individual's food culture and his goal is to 'work' with the diners to pique their interest in eating species other than the ones they already know.

While we were talking, Davide began to clean the fish and the cephalopods, which he always does with seawater. He showed us how he cuts up cuttlefish and squid. We noticed that he did not wash the ink off the delicate white meat from the mantles and he explained that the black was a natural element of the cephalopods. When we later saw one of his dishes we discovered that the ink also served an esthetic purpose. It created a fine contrast between the black surface and the inner white side of the cooked meat.

The kitchen opens out toward the restaurant, which is actually a covered terrace near the edge of the cliff that goes down to the sea, so that there is a wonderful view out over the water. We saw Davide and Guiseppe out on the cliff foraging for wild plants for the lunch. While we waited for our food, Mario commented on the presence in the distance of a row of trawlers that had been sailing back and forth all day. He pointed out that they were in international waters and scoop up most of the fish, leaving very little for the inshore fishers in their small boats.

When lunch was finally ready we started with *bottarga* accompanied by dandelion flowers and beach fennel. Davide had made the *bottarga* himself from the roe sacs of mullets that were caught in the lagoon. Next we were served strips of raw *Loligo* with julienned Sardinian artichokes, a little olive oil, and wood sorrel flowers. This was followed by the obligatory pasta dish, the traditional local *fregula*, which closely resembles couscous, topped with meaty common cockles from the lagoon. The main dish consisted of steamed and then grilled *Sepia* mantle with grilled nidamental glands, spinach, and wild sautéed asparagus, which were already in season locally in February. Because Davide had made cross-hatched cuts in the outside of the mantle it had curled up and looked like a small porcupine with lightly browned tips that resembled quills and a dark surface with cuts that opened up to reveal the chalk white interior. Desert was a white chocolate meringue with pepper, white chocolate mousse, and a foamy yogurt cream.

By the time we finished it was almost five o'clock and we had to drive back to Cagliari, where yet another cephalopod meal awaited us. Roberto had devised a test for his old friend, Pier Luigi Fais. Would he be able to make a complete meal with only cephalopods and was he ready to experiment and break away from the traditional Sardinian recipes? We knew that Pier Luigi had tried to make Roberto reveal what we had been doing with the cephalopod innards in Copenhagen. But Roberto had not let out any secrets in order to torment his culinary friends just a little.

Pier Luigi's restaurant, called Josto, is located in a former two-story warehouse with very high ceilings. The small interior has a modern post-industrial look, and is dominated by the view of the open kitchen. An interesting twist is the location of the 'wine cellar' on the main floor and from our table we could see a tempting array of bottles behind the glass panes. Background music came from an old-fashioned gramophone which was tended by one of the servers who put on a new record from time to time, choosing from a selection on a shelf on the wall.

In the kitchen, Pier Luigi and the second chef, Matteo Russo, together with a handful of helpers were busy preparing meals for us and for the other diners. From where we were sitting we could easily follow the intense effort required to cut up the mantle of a cuttlefish. We were filled with the joy of anticipation and curiosity about the nature of the courses that we were about to sample.

We started with raw pieces of *Loligo*, folded like rose petals around a slice of *lardo* with a granita made from a special Sardinian citrus fruit, *pompia*. It was a risky, but successful, proposition to combine fatback with the lean squid and counterbalance their tastes with the bitter and sour citrus notes. The texture of both the meats was soft and creamy, each in its own way. The second dish consisted of deep-fried *Sepia* arms, a crisp *puntar-ella* (chicory) salad, apple juice, and a lemon emulsion. The meat of the arms was soft and the batter was crisp. Next we were served strips of steamed cuttlefish mantle with charcoal grilled romaine. The crowning touch on this dish was a sauce made with wine vinegar and butter, seasoned with squid liver, which has abundant umami and a slight fish taste. We later found ourselves in agreement that this sauce was the high point of the meal and clearly demonstrated the way in which cephalopod liver can raise the taste of a dish to a whole new level.

A soup was next on the menu. Pier Luigi made a warm bouillon from octopus cooking water in which paper thin strips of *Loligo* mantle were cooked for ten minutes and seasoned with ginger and honey. All of us at the table agreed that this soup was exceptionally delicious and was yet another memorable aspect of the meal.

Before serving the next dish Pier Luigi asked us to come over to the counter that opened onto the kitchen to taste the octopus cooking water. It was very important to him to have the master chef, Roberto, deliver a verdict on the bouillon and to talk a little about the work of 'The Squid Squad' to use *Loligo* in untraditional ways. Soon five chefs were gathered around Roberto to learn something new. Tasting spoons were brought out to sample the cooking water and then Pier Luigi showed off steel containers with the various parts of the innards of *Loligo* and *Sepia*.

Using tweezers, Roberto pulled out a variety of organs from the coal black mass of innards. He gave a small lesson about each of them and how they could be used. From a distance this could have looked like a group of surgeons discussing the outcome of an operation. A spirited exchange of questions and ideas ensued. It became obvious to us that these Sardinian chefs were eager to view the raw ingredient 'cephalopod' as something that could be used in ways other than the classical ones with which they were familiar. It was interesting for us visitors to be in a position to impart new knowledge to them about a raw ingredient that is virtually unknown in Nordic cuisine. But by way of our project we had had the freedom to tackle cephalopod cuisine without being bound by any entrenched gastronomic traditions.

As could be expected there was no way around having a pasta dish in an Italian restaurant. In this case it was spaghetti with almonds, topped with grated cephalopod-*bushi*, that is, cephalopod prepared like a Japanese *katsuobushi*. This was surprising and innovative. It was made from *Loligo* fins that Pier Luigi had placed in a smoker overnight to dry them out and imbue them with a smoky smell and taste from the slow-burning charcoal.

Octopus had another star turn in the next dish in the form of dumplings made with potatoes and cooked octopus, served in a purée made from the leaves of wild beets. It was a simple, attractive looking dish. The final cephalopod dish was a return to the classics in the form of sautéed and simmered octopus arms, served with mashed potatoes and sautéed wild chicory. The octopus pieces were perfectly tender, creamy, and not the least bit dry. We finished the meal with a dessert combination that was totally devoid of cephalopods—ice cream with caramel sauce and a crumble topping together with a carrot sorbet.

Pier Luigi had clearly passed Roberto's test and we had the fantastic experience of eating a whole meal composed of different types of cephalopods without losing our enthusiasm for finding them in every single dish. The culinary team at Josto had shown that they were able to take a fresh look at what was for them a familiar raw ingredient, even though this had proved to be a real challenge. But a challenge brings change in its wake and the chefs from the restaurant and from the laboratory in Copenhagen had shared their knowledge and expertise. As we said goodbye to the very dedicated kitchen crew we invited them to make a visit to Denmark in return.

Sustainability in the Anthropocene Epoch—A Special Role for Cephalopods

According to projections, there will be almost ten billion people living on Earth by 2050. This enormous population increase will inevitably lead to a dramatic escalation in the global competition for natural resources and there will be a greater focus on the sustainability of the food supply from all points of view—economic, social, and environmental.

By now it has become more and more obvious that we are living in the Anthropocene epoch where we are making an irreversible impression on our surroundings. Virtually all life forms and even the inanimate, physical nature of the planet itself bear witness to the impact of human activities. The ways in which these are interrelated are complex and difficult, or even impossible, to understand fully and this has always been the case. What is new, however, is that our effects on the planet are no longer just local and temporary, but global and probably cannot be undone.

There is powerful evidence that climate change is anthropogenic and that both land and marine environments are already under great stress. This gives rise to the question of how we can share and administer the world's resources in a more sustainable fashion. But there is no unanimity about what constitutes sustainability nor about how resources should be allocated, both issues that are political and ethical minefields.

When it comes to meeting the challenge of feeding so many people, it is above all a question of ensuring that there is enough food to meet the minimum requirements for calories, proteins, essential fats, and other nutrients. It is difficult to imagine that there will be food security unless we take a two-pronged approach to the problem, considering both plant and animal sources in an integrated way. Supporting the on-going technological evolution in the agricultural sector to increase crop yields is part of the solution. But we need to recognize that intense cultivation entails the use of nitrogen and phosphorus fertilizers, as well as comprehensive pest control, both of which may work against attempts to mitigate agriculture's sometimes detrimental environmental footprint. And we also need to reassess whether reducing the overall amount of plant protein suitable for human consumption, such as soybeans, by making it into animal fodder is the most effective use of that resource.

In parallel, it is important to rethink how we can satisfy the still growing global demand for animal protein. Some forms of animal husbandry place an enormous burden on the environment. Cattle ranching, especially, results in greenhouse gas emissions that are fifteen to twenty times greater than those caused by the production of other protein sources such as insects, farmed salmon, and chicken. Consequently, one way to address the problem of how to meet the demand for animal protein is to reduce the production of meat from terrestrial animals and look to the oceans to provide us with more, and more sustainable, protein-rich food. This is where cephalopods come into the picture as a resource that is worth a closer look.

Food from the Sea for a Hungry World

The oceans are, in many ways, an inferior and, in many cases, an inappropriately exploited source of nutrition. With respect to fishing, humans are to a large extent still hunter gatherers. Wild fish stocks of many species have been depleted by overfishing and are at risk. While fish farming can partially compensate for declining catches, the problems posed by marine pollution associated with certain types of fish and shellfish aquaculture are imposing stricter limitations on that industry. One way to maximize our utilization of marine resources is, therefore, to diversify our diet by learning to eat species that are unfamiliar to us and also by seeking out ones that are lower down in the food chain. Furthermore, we must be prepared to eat more seafood in its original form rather than turning it into fodder for other animals destined for human consumption, a process that is analogous to feeding livestock with plant protein. The loss of valuable nutrients is just too great, estimated at more than 90 percent each time a raw ingredient is fed to an animal that is higher up in the food chain.

Seaweeds and algae make up a large proportion of the lower end of the marine food network and they are, among others, a source of valuable polyunsaturated fats. At present, there is an increasing emphasis on eating seaweeds. Moreover, fish and shellfish that have not traditionally been a part of many food cultures are becoming more widely accepted. In some cases, these could be smaller, unusual, or less-valued finfish species such as sprats, porgy, spiny dogfish, and haddock. But this idea could just as easily apply to molluscs and, in particular, to cephalopods.

Overfishing of different species indicates that we ought to eat more cephalopods, as until now they have exhibited much more ability than fish to adapt to the changing conditions in the oceans, including the significant rise in the water's temperature and acidity level attributed to climate change. In addition, cephalopods are able to convert their food intake to musculature very efficiently, grow quickly, and have a short life span. But trying to raise cephalopods in an aquacultural setting is notoriously difficult and this poses a challenge in exploiting them more fully as a sustainable food source.

Octopus Aquaculture—Pros and Cons

In some respects, octopuses are the more obvious candidates for cephalopod aquaculture because, unlike most cuttlefish and squid, they stay on the seabed and move around only over short distances. In addition, they can be sold for a good price. Consequently, the race is on to find ways to farm octopuses on a commercial scale. Attempts to do so are actively underway in Europe, Asia, Australia, and Latin America and, as of 2020, a number of these projects appear to be closing in on the prize.

The major problem in raising octopus in captivity is related to the life cycle of some species. Their tiny newly hatched offspring, known as *paralarvae*, swim around like plankton for two or three months before they settle on a habitat. At this stage they are critically dependent on getting the right food, which is normally a varied diet of live microcrustaceans and shellfish larvae. A popular fodder used in aquaculture, briny shrimp (*Artemia salina*), is sometimes used as a substitute, but this still poses some challenges. First of all, as briny shrimp are in high demand for many types of aquaculture and as aquarium pet food, the supply chain is not secure, and they may prove too expensive. Secondly, studies have shown that the *paralarvae* do not thrive properly if fed only briny shrimp as they lack some of the essential nutrients found in their natural, more diverse diet. In order to compensate for this, nutritional supplements must be added. Again, this may be a cost factor. Over and above this, the *paralarvae* are very sensitive to small fluctuations in temperature, dissolved oxygen levels, and salinity. These constraints are reflected in the variety of approaches that have been used to raise octopuses in captivity, a few of which are described below.

Groups in Spain and Japan have been working on a full-cycle aquaculture process for *Octopus vulgaris*, meaning that the new creatures are derived from the eggs of animals that were themselves conceived by artificial incubation. As the female octopus tends her eggs very carefully after they are fertilized, this breakthrough can have important consequences for survival rates. With carefully controlled conditions, new tank setups, and improved rearing techniques, they have been able to achieve survival rates of 50 percent or more. This is an incredible increase over the estimated 1 to 2 percent of hatchlings that grow to full size in the wild. In 2018 the Japanese seafood company Nissui announced that it had succeeded in hatching over 100,000 octopus eggs and that it hoped to start marketing its farmed octopus in 2020. Similarly, after decades of research, the Nueva Pescanova Group in Galicia, northern Spain, announced in 2019 that octopuses born in its installation had reached adulthood and started to reproduce themselves. This enterprise hopes to start selling octopus commercially in 2023.

In Mexico, it has proven feasible to farm a certain species of octopus, *Octopus maya*, which does not have a planktonic phase. While baby octopuses have been hatched from eggs and raised to maturity, this project has run into a number of problems. One is the need to hand-feed a squid and crab mixture to these tiny creatures, an extremely labour-intensive process, another is the difficulty in harvesting them *en masse* and killing them humanely, and yet another has been the difficulty in raising capital for the venture. There is still a long way to go before *Octopus maya* can be reared commercially. And even if the enterprise ultimately succeeds, this species may not be nearly as marketable as the very popular *Octopus vulgaris*.

As commercial octopus aquaculture comes closer and closer to becoming a full-blown industry, serious concerns remain about the very concept itself. Many of these were addressed in 2019 in an open letter signed by more than 100 scholars. Some of the issues are the following. Farmed octopuses are likely to be sold in up-scale markets in countries that already have a high degree of food security, so they will not contribute materially to feeding those who suffer from hunger. Because octopuses are carnivores, they must be fed with marine species that are lower down on the food web, which will increase rather than ease pressure on stocks of wild aquatic animals. On both counts they consider octopus aquaculture to be environmentally inefficient. There are also ethical reasons for having reservations that relate to the nature of octopuses as sentient beings. They seek out a high degree of cognitive stimulation and are generally asocial preferring to avoid others of the same species. Apart from that, octopuses are cannibals and will prey on each other. Given these characteristics, they are unlikely to adapt well to the tightly controlled, crowded, and monotonous conditions that are typical of aquaculture settings.

Ultimately the future of octopus farming may be driven not only by the need to expand the food supply but also determined by an increased understanding of the complex nature of octopuses.

Working Out the Life Cycle of Cuttlefish at Roscoff, France

The Station Biologique de Roscoff is an important marine biology research station located in Roscoff, at the northern-most point of Brittany facing the English Channel. It was founded in 1872 by Henri de Lacaze-Duthier (1821–1901), a pioneer in the study of living animals in their habitat. It is said that there were three principal reasons for his choice of this site. First of all, both the animal and plant life along the shore is exceptionally diverse. Secondly, the tide is low at midday, which maximizes the opportunity for scientists to work in the intertidal zone. Thirdly, the railroad had recently been extended to the area, which made travel to and from Paris much easier.

A few years ago in July, one of us (Ole) was able to catch up with Dr. Xavier Bailly, a developmental biologist associated with this research facility. He is one of the first people in the world to have reproduced successfully the life cycle of the cuttlefish *Sepia officinalis* in a laboratory setting.

As Xavier stressed repeatedly in the course of our conversation, one can hope to understand the biology of an animal only if one can reproduce its life cycle completely under controlled conditions in a laboratory. In the case of *Sepia*, the life cycle is two years. When I visited him, he was engaged in raising the second generation, that is, the offspring of cuttlefish that had themselves been raised in the laboratory from eggs laid in the wild and transferred to their fish tanks. The sixty or so small individuals we looked at in an aquarium had hatched about six weeks previously and had grown to the size of a fingernail.

◻ Station Biologique de Roscoff and Dr. Xavier Bailly showing his aquaculture installation for *Sepia officinalis*.

The tiny creatures were actively swimming around in the water, but they were actually in a critical phase of their life cycle where they will thrive only if they are fed with live small shrimp. This is a fundamental problem that is common to the aquaculture of cephalopods. Another important aspect is the quality of the seawater in the tanks and whether it has the right types of microflora and microfauna. Xavier's research has indicated, however, that once the hatchlings are a little older he can get them to eat frozen shrimp, which are much cheaper and easier to obtain. He was quite confident that it would be possible for him to end up with a generation of at least twenty fully grown cuttlefish from this batch.

Xavier had presented some live cuttlefish that were only six months old and about five to ten centimetres long to a group of top chefs. They were completely taken with these small cephalopods, which can be eaten whole after the beak and cuttlebone have been removed. According to the chefs these cuttlefish had a much more interesting taste and texture than adult ones. The catch of *Sepia* in the wild is subject to restrictions with regard to minimum size in some countries, which is reflected in how they are caught, for example, the size of the fish net mesh. As a result, juvenile cuttlefish with gourmet potential are not generally available commercially.

After talking with the chefs, Xavier became convinced that he could master the challenges of raising cuttlefish in aquaculture for this market. In order to do so sustainably, Xavier's laboratory will have to come up with a way to nourish the hatchlings with something other than live shrimp. At this point there is no clear answer to the question of whether the cuttlefish eat live shrimp because their movement attracts them or whether they have the right taste and texture. More research is needed to determine whether it will be possible to raise these juvenile cuttlefish on a scale that will elevate them to the status of a delicacy on restaurant menus.

Later in the year, at the beginning of November, I again visited Xavier's laboratory. The generation of *Sepia* I had seen earlier was thriving and they had grown to a size of five to ten centimetres. The outlook was very promising.

In 2015 the United Nations adopted the *2030 Agenda for Sustainable Development*, which sets out the need to take an ecosystem approach to fisheries and aquaculture. It calls for planning, development, and management of these sectors in a way that also takes into account not only their ecological impact, but also the social and economic aspects of sustainability. Fisheries and aquaculture are seen as crucial to improving food security and human nutrition. As one might expect, cephalopods are included in this framework.

It is, nevertheless, difficult to formulate plans for, or to control, cephalopod fisheries for two main reasons. One is that many of them are caught in international waters, where there are no enforceable quotas, and the other is that we still know so little about how these creatures live and thrive. There is very limited data that could be used to make credible predictions. Matters are further complicated by the very short life cycle of cephalopods.

Recent observations have, however, pointed to some surprising results that could indicate that there has been an increase in the cephalopod populations in all parts of the world over a relatively long period of time.

Are Cephalopods Really the 'Weeds of the Sea'?

Cephalopods have been called 'the weeds of the sea' because, like unwanted plants, their numbers are increasing, not only rapidly but in all parts of the oceans, even after the environmental conditions have changed. As suggested by a recent study there may be something to this rather dismissive characterization.

By analyzing data for the populations of thirty-five species of six different genuses (30 percent octopuses, 52 percent squid, 17 percent cuttlefish) covering the sixty-year period from 1953 to 2013 researchers observed clear evidence of growth in all groups. What was special about the selection of species for study was that it included ones that live at different depths of the sea, both those that are fished and those that are not harvested. In addition, some of the species were very locally bound, moving around within a radius of only a few kilometres in the course of their lives, while others included in the study travelled for thousands of kilometres through the world's oceans.

It can often be difficult to evaluate marine populations based on the reported catches, but those for cephalopods have shown an upward tendency. Given that the population data showed an identical pattern both for species that are a food source and those that are not, it is possible to conclude that the increase in catch sizes is not due to improved fishing techniques. These investigations demonstrate that the present catch rate is not causing a decline in the cephalopod populations—quite the opposite. So it is necessary to look for an underlying cause for the way in which the cephalopod populations are developing that is common to all of them.

The results of this study have been interpreted as demonstrating that the cephalopods have benefitted from anthropogenic climate change. It is known that these creatures react very quickly to changes in the environment, especially water temperature. Climate change has led to warmer seas and some species have migrated northward. What is not as yet known is how this will affect both the cephalopods themselves and the other species that inhabit these colder waters.

It is possible that a decrease in fish stocks may have had two seemingly opposite effects. The decline in the fish populations that compete with and prey on the cephalopods may have led to better living conditions for the latter. And, conversely, there are also fewer of those fish on which they themselves depend for food. Both of these scenarios go right back to the prehistoric evolutionary competition between finfish and cephalopods and the equilibrium between these two very different marine life forms. At one time, about 100 million years ago, the cephalopods ruled the seas. But they lost out to other creatures, among them the fish, and their species diversity was reduced dramatically. Only about 800 species of cephalopods have survived to the present day, while there are some 30,000 species of finfish.

Perhaps this imbalance will soon be redressed if the very evident growth in cephalopod populations in the course of the past sixty years continues. A possible explanation is that the combination of very efficient physical growth patterns, short life cycles, and a consequent ability to adapt to rapidly changing conditions might work in their favour. It simply leaves them in a better position to accommodate themselves to sudden changes in the food chain and their physical surroundings than organisms that live longer and mature more slowly. This may help the cephalopods, once again, to dominate the marine environment.

A Special Place for Cephalopods on the Menu

This book has focused on demonstrating that cephalopods are not just fascinating creatures, but are also a wonderful, sustainable food source thanks to their physiological attributes, wide geographical distribution, and special tastes. They live in all the oceans, are easy to work with, and can be turned into a vast range of delicious dishes.

Moreover, cephalopods may have a very special role to play in weaning people away from consuming so much meat. The 2019 report from the EAT-Lancet Commission on Healthy Diets has documented that the conditions for a sustainable, healthy, and nutritious diet for the growing global population can only be brought about by an urgently needed and major change in the global food systems. This will require switching to a much more plant-based food intake and reducing red meat consumption drastically. The question arises as to how this can be done.

The barriers to eating enough vegetables and fruit may be psychological, physiological, cultural, and social. It is important to acknowledge that plant-based food is lacking in umami, the basic taste that humans have evolved to crave. While sweetness, another much desired taste, is found in most fruits, it is virtually absent in vegetables. Without confronting these fundamental facts, we may fail to achieve the goal of a sustainable future for humanity. A possible solution to this challenge is to adopt a holistic flexitarian approach shaped by fundamental insights into taste and, in particular, how to introduce umami into a green diet. This would entail giving vegetables the top billing and using meat more like a condiment and a taste enhancer. Given its qualities as a nutritious and palatable food source, cephalopod meat should be viewed in a new light as an underutilized food resource—one that appears able to thrive in spite of the impact of human activities on the planet. When consumed in small, sustainable quantities, octopuses, squid, and cuttlefish could act as 'umamifiers' on a plate of vegetables that would make it more appealing for flexitarians.

Nevertheless, there are many elements that need to be factored in when considering whether and how to integrate cephalopods more widely into the food system. These have to do with whether we think they are animals similar to those we already eat, the extent to which they will fit into some food cultures, and even gender related diet patterns.

In England cephalopods enjoy the same animal rights as vertebrates when it comes to animal experimentation and the European Union seems to be moving in the same direction. *The Cambridge Declaration on Consciousness,* which dates from 2012, goes as far as to single out octopuses explicitly as animals that, like birds and mammals, have a sufficiently complex neurological structure to have the basis for consciousness. In this way, octopuses are considered to be something like honorary mammals.

The Cambridge Declaration on Consciousness: "The absence of a neocortex does not appear to preclude an organism from experiencing affective states. Convergent evidence indicates that non-human animals have the neuroanatomical, neurochemical, and neurophysiological substrates of conscious states along with the capacity to exhibit intentional behaviours. Consequently, the weight of evidence indicates that humans are not unique in possessing the neurological substrates that generate consciousness. Nonhuman animals, including all mammals and birds, and many other creatures, including octopuses, also possess these neurological substrates."

Cephalopods constitute a valuable alternative protein source for those who absolutely want to eat meat or who may feel vaguely guilty about the carbon footprint associated with animal husbandry. But as it is becoming clear that some of the cephalopods, most notably octopuses, have a very advanced neural system and a brain that may be the seat of consciousness, much like vertebrates, ethical considerations might cause others to avoid them. So there is no escaping the question of whether we should be eating them at all. The American philosopher Michael Tye discusses this existential problem in his 2016 book, *Tense Bees and Shell-Shocked Crabs: Are Animals Conscious?* He argues that we should not necessarily focus on consciousness to decide whether or not we can eat certain creatures. Nevertheless, given their behaviours and obvious signs of intelligence, we should admire them, and treat them humanely with respect, even if they do not have a neocortex like us. One indication of a move to adopt this philosophy relates to the sometimes brutal ways in which cephalopods are handled by fishers. Steps have recently been taken to develop methods for humane slaughtering of cephalopods as is the case for higher animals used for food.

There are also a number of stumbling blocks that stand in the way of gaining a wider acceptance for cephalopods in some food cultures. Where they and other 'exotic' ingredients have not already found a place in a traditional cuisine, these visually different creatures may not be greeted with great enthusiasm. We are well aware that, in the long run, it takes more to alter eating habits than information about sustainability, a healthy diet, and respect for the natural world. It requires a cultural shift to become accustomed to new foodstuffs and their particular tastes. And if they are eaten more as a novelty than because they are delicious, they will soon fall out of favor. In order to exploit the full potential of cephalopods in those parts of the world, it will be necessary to make catches more readily available at the level of the individual consumers, so that they can gain a much better appreciation of how satisfying it is to enjoy them when they are freshly caught.

Gender differences also play a role in the further move towards more sustainable eating involving cephalopods. Women appear to embrace a more plant-based diet that also incorporates seaweeds and nutritional microalgae more readily than men. This is consistent with the observation that meat is considered a masculine form of food. Cephalopods may, however, come to the rescue since they are meat and can, in small amounts, substitute for that particular craving.

In the opinion of the authors of this book, sustainable eating has to be built on a holistic approach. Asking a significant part of the global population to forgo animal products and become vegetarians or vegans is simply not a realistic option, neither in the short term nor in the long term. Sustainable eating on a global scale is not a one way street and cephalopods have a role to play here.

There is much that we do not know about cephalopods and there are still many enduring myths that should either be shown to have validity or be stamped out. In their classical work *Cephalopod Behaviour*, two well-known cephalopod researchers, Roger T. Hanlon and John B. Messenger, write that "it is sad to reflect that the truth as it emerges will almost certainly be less exciting than the fiction, at least to some people."

But it is not certain that this observation will prove to be the correct one when it comes to the possibility that these fantastic creatures possess an intelligence and a mind that puts that of humans in a new perspective. Nevertheless, it is equally plausible that cephalopods caught in the wild, as well as those raised in aquaculture, can become an important component of the food of the future and help to ensure that we will be able to feed a growing world population.

■ Fossil of a nautilus (Nautiloidea), circa 175 million years old, found in Brittany.

The Technical Details

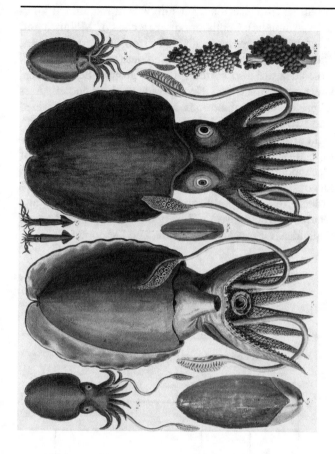

The Cephalopod Family Tree

Mollusca (molluscs) is a phylum within the animal kingdom (Animalia). Cephalopods are a class of Mollusca and they are themselves divided into two sub-classes, Nautiloidea (nautiluses) and Coleoidea (true cephalopods). Coleoidea are also sorted into two groups, Belemnoidea, which are extinct, and Neocoleoidea, which encompasses all present-day cephalopods. Neocoleoidea are also divided into two superorders, Decapodiformes (cephalopods with eight arms and two tentacles) and Octopodiformes (cephalopods with eight arms). There are four orders of Decapodiformes: Spirulida (extinct), Sepioidea (cuttlefish), Oegopsida (squid), and Myopsida (squid). There are two orders of Octopodiformes: Vampyromorphida, and Octopoda (octopuses). This book deals primarily with Sepioidea, Oegopsida, Myopsida, and Octopoda.

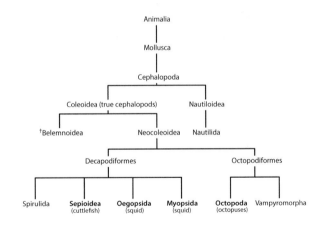

Cephalopods Included in This Book

The table lists the cephalopods that are discussed in this book according to the name of their species, together with the common English names. In addition, there is a notation of whether a species can, in principle, be used for human consumption, even though it may not be common to do so or to catch it. The right hand column indicates where that species is usually found.

Latin name		English name	Edible?	Primary distribution
Alloteuthis subulata	S	European common squid	Yes	Eastern North Atlantic Ocean
Architeuthis dux	S	North Atlantic squid	No	All oceans, but rare in tropical and polar latitudes
Argonauta sp.		Argonaut, paper nautilus	Yes	Tropical and subtropical waters worldwide
Callistoctopus macropus	O	White-spotted octopus	Yes	Mediterranean Sea, warmer parts of Atlantic Ocean, Caribbean Sea, Indo-Pacific Ocean
Dosidicus gigas	S	Jumbo flying squid, Humboldt squid	Yes	Eastern Pacific Ocean from Tierra del Fuego to California
Eledone cirrhosa	O	Curled octopus, horned octopus	Yes	Eastern North Atlantic Ocean, Mediterranean Sea
Enteroctopus dofleini	O	North Pacific giant octopus	Yes	Coastal North Pacific Ocean
Euprymna scolopes	C	Hawaiian bobtail squid	No	Pacific Ocean
Grimpoteuthis sp.	O	Dumbo octopus	No	All oceans, but at great depths
Hapalochlaena spp.	O	Southern blue-ringed octopus	No	Indo-Pacific Ocean (blue- ringed octopus are found as far north as Japan)
Idiosepius thailandicus	C	Bobtail squid, pygmy squid	No	Indo-Pacific Ocean
Illex argentinus	S	Argentine shortfin squid	Yes	Southwestern Atlantic Ocean
Loligo forbesii	S	Veined squid, long-finned squid	Yes	Eastern North Atlantic Ocean, Mediterranean Sea, Red Sea, Western Indian Ocean

Latin name		English name	Edible?	Primary distribution
Loligo pealei	S	Longfin inshore squid	Yes	North Atlantic Ocean
Loligo vulgaris	S	European squid	Yes	Coastal waters of the eastern North Atlantic Ocean
Mesonychoteuthis hamiltoni	S	Colossal squid, Antarctic squid, giant cranch squid	Yes	All parts of the Southern Ocean
Nautilus sp.		Nautilus	Yes	Indo-Pacific Ocean
Octopus maya	O	Mexican four-eyed octopus	Yes	Tropical Western Atlantic Ocean
Octopus vulgaris	O	Common octopus	Yes	Atlantic Ocean, Indian Ocean, Pacific Ocean
Octopus wolfi	O	Star-sucker pygmy octopus	No	Western Pacific Ocean
Sepia officinalis	C	Common cuttlefish, European common cuttlefish	Yes	Mediterranean Sea, North Sea, Baltic Sea
Sepietta oweniana	C	Common bobtail	Yes	Eastern North Atlantic Ocean, Mediterranean Sea
Sepiola spp.	C	Bobtail squid	Yes	All oceans
Taningia danae	S	Dana octopus squid, Taning's octopus squid	Yes	Atlantic Ocean, Pacific Ocean, but very rare
Todarodes pacificus	S	Japanese flying squid, Japanese common squid, Pacific flying squid	Yes	Northern Pacific Ocean
Todarodes sagittatus	S	European flying squid	Yes	Northeastern Atlantic Ocean, Mediterranean Sea
Thaumoctopus mimicus	O	Mimic octopus	No	Indo-Pacific Ocean
Watasenia scintillans	S	Firefly squid, sparkling enope squid	Yes	Western Pacific Ocean

C cuttlefish, O octopus, S squid

Glossary

ambergris - a special secretion that is formed in the intestines of the sperm whale in the shape of grey rock-like lumps that can weigh up to several hundred kilograms and often contain the remains of cephalopods; known since ancient times as a fixative in perfumes and for its fragrance.

Animalia - the animal kingdom.

anisakis - parasitic nematode (herring worm) that can infest certain fish and cephalopods.

adenosine triphosphate - (adenosine-5'-triphosphate, ATP) polynucleotide that is the biochemical source of energy production in living cells. Among other substances, it can be broken down to the 5'-ribonucleotides inosinate, adenylate, and guanylate, which are associated with synergistic umami.

adenosine monophosphate - (AMP), a salt of the nucleic acid adenylic acid; synergizes with glutamate to enhance umami; found especially in fish, shellfish, and octopuses.

alcatruz - pot used to catch octopuses in the traditional way by Mediterranean fishers.

amino acid - amino acids are the fundamental building blocks of proteins. Examples include glycine, glutamic acid, alanine, proline, and arginine. Nature makes use of 20 different, specific amino acids to construct proteins, which are chains of amino acids bound together with peptide bonds. Short chains are called polypeptides and long ones proteins. In food, amino acids are often found bound together in proteins and also as free amino acids that can have an effect on taste. An example is glutamic acid, which is the basis of umami. More than half of the free amino acids in cephalopods are of the type that are associated with the taste of seafood (arginine, glutamic acid, alanine, and glycine).

ammonium chloride - sal ammonium or salmiac (NH_4Cl) found throughout the mantle of the giant squid *Architeuthis dux*, as well as in other cephalopods that live at great depths. As it is lighter than seawater it helps to regulate buoyancy in the animals.

Anthropocene epoch - new geological epoch dating from the mid-1950s, the point at which humanity began to make irrevocable impressions on the Earth's ecosystems.

Architeuthis - from Greek *archē* (chief or principal) and *teuthis* (squid); genus name for the giant squid *Architeuthis dux*.

argonaut - (paper nautilus) cephalopod belonging to the species *Argonauta*, enclosed in a shell.

arthropods - phylum of invertebrates with exoskeletons, for example, insects and crustaceans.

ATP - see adenosine triphosphate.

axon - long, thin projection of a nerve cell that sends signals to other nerve cells.

belemnites - (Belemnoidea) group of cephalopods that were abundant during the Cretaceous Period (from about 145 million to 66 million years ago) and that became extinct at the same time as dinosaurs; fossilized remains are known popularly as thunderstones.

bilateral organism - organism that has two virtually identical symmetrical sides, for example, cephalopods and humans.

biogenetic amines - substances that can be formed by an organism's own enzymes and by microbial activity; potentially toxic when present in large quantities.

bioluminescence - active light emitted by some organisms; produced when certain chemical reactions take place in bacteria that live in symbiosis with the host organism.

blue-ringed octopuses - *Hapalochlaena*, genus of four different species of highly venomous octopuses found in the Indo-Pacific region.

bromelain - enzyme derived from fresh pineapples that can break down proteins such as collagen and gelatine, used as a meat tenderizer.

calamar - also calamari, calamares, common names for various types of squid.

Cambrian Explosion - event about 542 million years ago and lasting for about 50 million years. During this period there was an amazing outburst of new life forms, among them all the major phyla of present-day animals, including the molluscs (Mollusca). Cephalopods emerged toward the end of this time.

Cephalopoda - from the Greek *kephalē* (head) and *pous* (foot); class within the invertebrate phylum (Mollusca) made up of about 800 still living species, divided into two sub-classes: true cephalopods (Coleoida) and nautiluses (Nautiloida).

chemoreceptor - receptor that can bind and identify a particular chemical substance.

chitin - polysaccharide that is a component of the cell walls of some fungi and the exoskeleton of insects and crustaceans. Cephalopod beaks and the gladius of squid are made of chitin.

chromatophores - small complex organs made up of pigment-containing cells that can reflect light.

cirri - appendages around the mouth of Nautilida (nautiluses).

class - biological classification between phylum and order, for example, Cephalopoda are a class under Mollusca.

Cnidaria - phylum of invertebrates, characterized by special cells used for capturing prey and bodies made up of a non-living jelly-like substance; divided into polyps, such as sea anemones, that are sessile and medusae, such as jellyfish, that swim.

Coleoidea - sub-class that groups together all the cephalopods that have no shells; divided into the Belemnoidea, now extinct, and the Neocoleoidea, that includes all present-day cephalopods. Their common characteristics are a number of arms surrounding the mouth, very highly developed eyes, three hearts, and the ability to eject ink.

collagen - protein network that forms connective tissues and thereby gives structure to all animal tissues, found mostly in the skin, muscles, and bones. The extent to which the collagen fibres are cross-linked determines how tender or tough a meat will be. It can be broken down to water-soluble gelatine by long-term heating at more than 70° Celsius for mammals and 60° Celsius for cephalopods.

colossal squid - also Antarctic squid or giant cranch squid (*Mesonychteuthis hamiltoni*), believed to be the world's largest squid species.

connective tissue - see collagen.

cross-linking - formation of crosswise chemical bonds between long-chained polymers, for example, proteins or carbohydrates. Both collagen in the skin and muscles and cellulose in the cell walls of plants are strongly cross-linked, which is why these tissues can be stiff and tough.

curled octopus - also called horned octopus, *Eledone cirrhosa.*

cuttlebone - a sort of inner shell in cuttlefish composed of calcium carbonate together with small amounts of mineral salts.

cuttlefish - cephalopod with eight arms and two tentacles; belongs to the order Sepioidea.

Decapodiformes - superorder of Neocoleoidea cephalopods with eight arms and two tentacles; encompasses four orders: Spirulida, Sepioidea (cuttlefish), Oegopsida (squid), and Myopsida (squid).

dendrite - projection from a nerve cell that receives signals from other nerve cells.

Devonian - geological period from 419 million to 359 million years ago.

DHA - docosahexaenoic acid, polyunsaturated fatty acid found in large quantities in marine organisms.

disruptive colouration - mechanism whereby an organism, for example an octopus, can blend into its surroundings by taking on colours and patterns that break up its normal contours.

dopamine - neurotransmitter; substance that sends signals between nerve cells and in the brain.

endocrine system - collection of glands that secrete hormones that integrate and control bodily metabolism.

EPA - eicosapentaenoic acid, polyunsaturated fatty acid found in large quantities in marine organisms.

epigenetics - discipline that studies how inheritable traits of an organism are changed by a modification of gene expression rather than by alterations in the DNA sequence.

eumelanin - see melanin.

European squid - *Loligo vulgaris.*

family - biological classification that is between order and genus.

gastrophysics - interdisciplinary science defined by quali-tative reflections and quantitative investigations of foods, how they are handled and prepared, as well as their taste, focusing on physical effects and explanations.

gastro-lab - a particular type of laboratory certified for food handling, but with much more specialized and pre-cise equipment than would be found in an ordinary kitchen.

gelatine - protein found in the form of collagen in con-nective tissue that is released when collagen is heated, dissolving the stiff collagen fibres. In contrast to colla-gen, gelatine is soluble in water. When gelatine is cooled, the stiff fibre structure of collagen is not reformed and, instead, a water-retaining gel is created.

genus - biological classification between family and species.

gladius - commonly called a pen; vestigial rudimentary support structure in a squid that resembles a sword or a feather, somewhat flexible and made of chitin

glutamate - salt of the amino acid glutamic acid, for example, monosodium glutamate (MSG). In water, glu-tamate splits into sodium ions and glutamate ions, the latter being the main source of umami.

glutamic acid - amino acid; in ionic form its salts (gluta-mates) are a source of umami.

glycogen - branched polysaccharide molecule made up of glucose units. Glycogen is the energy depot in the liver and in the white muscles of fish, shellfish, and cephalo-pods.

Gorgons - in Greek mythology, three female-like beings, of which Medusa is the most famous; usually depicted as hideous monsters with hair made of living, writhing snakes.

gyotaku - traditional Japanese technique for making a print of a fish by pressing rice paper onto a fish that has been covered with ink.

hectocotylus - from Latin *hecto* (hundred) and Greek *kotúlē* (small cup), modified arm of a cephalopod with a particular organ (ligula) at the end that can transfer a protein capsule containing sperm (spermatozoa) to the female during mating.

hemocyanin - protein that transports oxygen in the bloodstream of invertebrates such as cephalopods and certain crustaceans, for example, lobsters. Instead of iron, hemocyanin relies on copper to bind the oxygen molecules, which accounts for the blue colour of cephalopod blood.

hemoglobin - protein that transports oxygen in the red blood cells of all vertebrates.

hepatopancreas - in fish and molluscs the glandular organ that combines the functions of liver and pancreas, secreting vital digestive enzymes. It is also responsible for the exchange of nutrients and waste products that are later expelled via the anus and for the flow of water out through the siphon.

hydrolysis - chemical process by which a molecule such as a protein is cleaved into smaller entities while absorbing water. For example, connective tissue (collagen) can be hydrolyzed to form gelatine.

hydrostat - a deformable system that is subject to constant pressure; the muscles of cephalopods are muscular hydrostats.

ikebana - centuries old traditional Japanese art of flower arranging.

invertebrate - animal that lacks a backbone or a bony skeleton; imprecise taxonomic classification for different animal groups that lack a spinal column, for example, insects, molluscs, worms, and crustaceans. Cephalopods are invertebrates.

iridophores - organs consisting of very thin transparent cells made of chitin sheets that can cause iridescence. By reflecting and refracting light from their surroundings they can produce metallic, silvery, blue, and green tones by a mechanism known as 'structural interference' in the same way as soap bubbles display a colour spectrum. Iridophores have the special property of being able to reflect polarized light.

iridescence - see iridophores.

Japanese flying squid - *Todarodes pacificus*, also called Pacific flying squid, belong to the Ommastrephidae family.

kingdom - in biology the second highest taxonomic rank divided between plants and animals.

Kraken - mythological giant cephalopod-like sea monster that can have either eight or ten arms. *Kraken* is a Norwegian word that has connotations of something with a frightening and twisted appearance.

leucophores - organs made up of static flat white cells that are able only to reinforce the effects created by chromatophores and iridophores; provide a chalk white background in the skin of cephalopods.

ligula - from Latin for small tongue; small cup-like organ on the hectocotylus used by male cephalopods to transfer sperm to the female during mating.

mantle - the muscular structure of cephalopods that encases the innards.

medusa - see Cnidaria.

melanin - dark biological pigment. There are several types of melanin, of which eumelanin is the most common one and is responsible, among other things, for the colour of cephalopod ink.

Mollusca - the phylum of molluscs.

mollusc - from Latin *mollis* (soft); invertebrate belonging to the phylum Mollusca.

MSG - monosodium glutamate; the sodium salt of glutamic acid; the most important source of umami.

Myopsida - order of squid that includes, among others, *Todarodes* spp. and *Architeuthis dux*; differentiated from the other order of squid, Oegopsida, by the structure of their eyes.

nautilus - cephalopod belonging to a sub-class of Mollusca, Nautiloidea.

Nautilida - order under Cephalopoda that includes cephalopods (nautiluses) with outer shells.

Nautiloidea - sub-class of Cephalopoda.

Neocoleoidea - group of all presently living cephalopods with eight arms (Octopodaformes) and cephalopods with eight arms and two tentacles (Decapodiformes).

neocortex - particular part of the mammalian brain that deals with the more complex functions such as cognition, sensory perception, planning, speech, and rational behaviour.

nessa - a net trap used by Mediterranean fishers for catching octopuses.

neurotransmitter - chemical substance that carries the signals between nerve cells, for example, dopamine.

nidamental glands - pair of glands in some species of female squid and cuttlefish that secrete the gelatinous substance that forms egg cases.

North Atlantic squid - *Architeuthis dux*.

nucleotide - chemical group that forms part of a nucleic acid, for example, ATP and the umami taste substance adenylate.

Octopoda - order of all present day octopuses.

Octopodiformes - superorder of Neocoleoidea cephalopods with eight arms; encompasses two orders: Vampyromorphida and Octopoda.

octopus - cephalopod with eight arms; belongs to the order Octopoda.

Oegopsida - order of squid that includes, among others, *Loligo* spp. and *Allotheuthis* spp.; differentiated from the other order of squid, Myopsida, by the structure of their eyes.

Ommastrephida - family of squid the includes the species *Illex argentius* and *Todarodes pacificus*, the most important commercially fished squid.

opsin - photoreceptor that responds to different colours of light; found in the skin of some cephalopods.

order - biological classification that is between class and family, for example, Octopoda.

oryza cystein protease - enzyme found in malted rice that can suppress the action of other enzymes.

paper argonaut - common name for a nautilus; refers to its fragile, translucent shell.

papillae - millimetre-sized protrusions in the skin of some cephalopods; controlled by special muscles. They can change the surface appearance of a normally smooth-skinned animal to make it look spiky or bumpy.

paralarva - stage in the life cycle of a newly hatched octopuses or squid when it swims around in the sea like plankton until it reaches the sub-adult stage and finds a permanent habitat.

patera - small net traps used in the Mediterranean for catching octopuses.

pen - see gladius.

peptide - molecule that is composed of amino acids, such as a protein.

Permian - geological period from 299 million to 251 million years ago.

peschiera - traditional Sardinian community co-operative that had stewardship over a piece of coastal land and the fishing rights associated with it.

photophore - photo-active organ that emits active light; in cephalopods this can take the form of bioluminescence due to chemical reactions that take place in certain bacteria that live symbiotically in the tissue of the animal.

photoreceptor - sense organ that responds to light falling on it.

phylum - broad biological classification just below kingdom and above class.

polarized light - light that is passed through a filter so that the waves of light are limited to one plane and oscillations are restricted to it. Unpolarized light, for example, from an incandescent lamp or the sun, oscillates in all directions.

polyp - see Cnidaria.

polysaccharide - carbohydrate consisting of many sugar units (saccharides).

protein - polypeptide; that is, a long chain of amino acids held together by peptide bonds. Enzymes are a particular class of proteins that act as catalysts for chemical reactions under controlled circumstances. Proteins lose their functionality (denature) and undergo changes to their physical characteristics when they are heated, exposed to salt or acid when cooked, salted, or marinated, or subjected to the action of enzymes when fermented.

potera - special net fish traps, often used to catch octopus.

protease - enzyme that can break down proteins.

pseudomorph - ink cloud with a greater mucus content that can hold its shape and resemble another cephalopod or a dangerous animal; used as a escape mechanism to confuse a predator.

radula - ribbon-like raspy tongue with rows of tiny, sharp chitin teeth in cephalopods.

Sepioidea - order of cephalopods to which cuttlefish and the closely related bobtail squid belong.

shunga - Japanese erotic art; usually in the form of a woodblock print.

siphon - muscular structure in the mantle of cephalopods through which the animal can expel water from the mantle into its surroundings; when done with great force this enables it to use jet propulsion.

Silurian - geological period from 444 million to 419 million years ago.

species - the basic unit of biological classification, for example, *Loligo pealei*, a species in the genus *Loligo*.

spermatozoa - small sac of male reproductive cells.

spinal chord - from Latin *chorda* (chord) and *spinae* (thorns), long chord of nerve tissue that is the major component of the vertebrate central nervous system.

Spirulida - order of cephalopods; with only one extant species, the ram's horn squid, which has a distinct coiled shell.

sp. - when written after a genus name, sp. indicates a reference to an unspecified species of the genus; for example *Octopus* sp.

spp. - when written after a genus name, spp. indicates that the reference is to several species within in the same genus; for example, *Hapalochlaena* spp. refers to all the species of blue-ringed octopus.

squid - cephalopod with eight arms and two tentacles; belongs to either the Myopsida or the Oegopsida order.

synapse - junction that permits a nerve cell to transmit a signal to an adjacent nerve cell.

Taishokan - Japanese legend from the Asuka period (538–710) that relates the story of the struggle between a young girl diver and a sea dragon-serpent (sometimes pictured as an octopus) to recover a precious jewel. Over the centuries the theme was sexualized and variations of it are found in Japanese art works that have subsequently inspired erotic depictions internationally in paintings, literature, and movies, often involving giant octopuses and squid.

teleost - infraclass of the ray-finned fish with a particular structure of the jaws.

tentacles - elongated appendages of squid and cuttlefish that are used to hunt prey.

tetrodotoxin - lethal venom in the bite of blue-ringed octopuses (*Hapalochlaena*); the same neurotoxin is found in pufferfish.

Teuthida - obsolete classification for the order to which squid belong; has been replaced by two separate orders, Myopsida and Oegopsida as they are now known to have descended from separate branches of the cephalopod evolutionary tree.

teuthis - ancient Greek word for a small squid.

thunderstone - popular name for the fossilized rostrum of a belemnite, a now extinct cephalopod. Early Europeans thought they were formed when thunderbolts hit the ground.

Triassic - geological period from 252 million years ago to 201 million years ago.

trimethylamine - foul-smelling organic substance (tertiary amine) produced, for example, by bacterial decomposition of trimethylaminoxide in dead seaweeds, cephalopods, and fish.

tyrosinase - enzyme involved in the synthesis of melanin and also a component of cephalopod ink. Its presence can cause eye irritation and confuse the sense of taste and smell of an attacking predator.

umami - one of the five basic tastes. There are two separate aspects of umami, a basal contribution, based on free glutamate, and a strengthening or synergistic contribution, which is due to the presence of 5'-ribonucleotides, especially inosinate, adenylate and guanylate.

Vampyromorphida - order (vampire squid) of cephalopods with only one extant species, which is characterized by an extended web that connects its arms.

Culinary Terms

agar - gelation agent extracted from red algae.

aliño - Andalusian sauce made with olive oil, wine vinegar (usually sherry vinegar), salt, green peppercorns, and finely chopped tomatoes and onions.

amok - (also *mok, ho mok*) in southeast Asian cuisine a curry that is steamed in a banana leaf, typically made with fish, galangal, and coconut cream and served with cooked rice. *Amok* is a Cambodian specialty.

azuki - small green or red beans (*Phaseolus angularis*). The red ones are sweet and used in the form of a paste in Japanese cakes, confections, and desserts.

bai-yo - Thai name for *Morinda citrifolia*, the leaves of which as used in Asian dishes.

bottarga - Mediterranean specialty of dried fish roe from tuna, cod, or mullet. The roe sacs are extracted and put in sea salt for a few weeks and then hung up to air-dry for about a month. The salt draws the liquid out of the roe, making it very firm and hard. It is rich in umami and served with tapas or cut into thin slices and used as a pasta topping.

brioche - delicate bun or bread made with a yeast dough that contains butter and eggs.

burrata - semi-soft fresh Italian cheese with an outer curd made from fresh buffalo milk mozzarella that is filled with stringy buffalo milk curds (*stracciatella*) and cream.

calamar en su tinto - cephalopod, usually cuttlefish, prepared in its own ink.

calamares a la plancha - Spanish expression for grilled squid or cuttlefish.

calamares chiquititos (*puntillitas*) - Spanish expression for the European common squid (*Alloteuthis sublata*).

calamares grelhados - Portuguese expression for grilled squid or cuttlefish.

calamaritos (*chipirones*) - Spanish name for European squid (*Loligo vulgaris*).

ceviche - raw fish, shellfish, or cephalopod meat that has been marinated in citrus juice, which makes it firmer and partly preserves it.

chipirones - Spanish name for European squid (*Loligo vulgaris*).

chipotle - pepper made from smoked, dried *jalapeños*.

chirashi-sushi - (Japanese for 'scattered sushi') a particularly colourful type of sushi in which fish, shellfish, and green vegetables are placed on a layer of sushi rice.

chocos fritos - popular Spanish expression for the deep-fried mantles of small *Sepia officinalis*.

chokkara - Korean name for an enzymatic fermentation medium made with the innards of cephalopods. See also *shiokara*.

daikon - (Japanese for 'big root') mild-tasting large, long white radish.

dashi - (Japanese for 'cooked extract') a stock made from, for example, seaweed (*konbu*) and bonito fish flakes (*katsuobushi*). *Dashi* is the epitome of an ingredient that adds umami.

emulsifier - a substance that depresses the surface tension between oil and water, thereby facilitating the formation of an emulsion. Lipids are emulsifiers.

emulsion - a mixture of water with oil-like substances, for example, fats, that are sparingly soluble in water. Mayonnaise and ice cream are examples of emulsions.

farinata - pancake made with chickpea flour; specialty of the Ligurian coast of Italy.

fregula - special type of Sardinian pasta made from semolina dough rolled into tiny spheres and baked in an oven.

galangal - any one of four species of aromatic rhizomes from the ginger family; used as a spice in Asian cuisine.

garum - brownish liquid that seeps out when salted small fish and fish innards from, for example, mackerel and tuna are crushed and fermented for a long time; similar to Asian fish sauces. *Garum* production was an important industry in ancient Rome and Greece.

glace - meat juice that has been reduced slowly; contains concentrated taste substances and can be used as a taste additive in gravies.

glaze - to give a smooth, glossy appearance by coating a surface with reduced aspic, butter, or sugar.

gnocchi - small round dumplings usually made from semolina, wheat flour, or potatoes; used much like pasta.

granita - sherbet prepared with a light sugar syrup, possibly with a little alcohol. Its texture is not uniform as it has small ice crystals that crunch between the teeth.

gremolata - condiment, typically made with chopped herbs, lemon zest, and garlic.

grissini - crisp bread sticks.

hashi - Japanese word for chopsticks.

hotara ika no shiokara - variant of the Japanese dish *ika no shiokara* made with small firefly squid (*Watasenia scintillans*).

huevos de choco - Andalusian specialty made with the nidamental glands from large female *Sepia officinalis*.

ichiya-boshi - Japanese expression for fish or cephalopods that are salted and dried overnight.

ika - Japanese word for cuttlefish and squid, of which at least 100 species are found in the waters around Japan. *Mongo-ika* are cuttlefish and *kensaki-ika* are squid.

ika no ichiya-boshi - Japanese expression for semi-dried squid.

ika no shiokara - squid or cuttlefish that is fermented in the enzymes from its own liver (hepatopancreas).

ika-sōmen - noodle-like strips cut from the mantle of a cuttlefish.

ishiru - fermented Japanese fish sauce that is high in glutamate.

jiaoyán yóujú - crispy fried squid seasoned with salt and Sichuan pepper.

julienne - description of food that is cut into thin strips like matchsticks.

Kampot pepper - cultivar of black pepper (*Piper nigrum*) grown in the Kampot province of Cambodia; available in green, black, red, white, and yellow varieties depending on how ripe it is when harvested and whether it has been dried.

katsuobushi - Japanese expression for a hard filet of *katsuo* (bonito) that has been subjected to a comprehensive process of cooking, drying, salting, smoking, and fermenting; contains large quantities of inosinate, which contributes synergy to umami.

kimchi - Korean name for pickled, fermented vegetables, for example, cabbage.

kokumi - Japanese expression that describes the continuity, that is, the long-lasting taste impression, and mouthfeel of food. There may be some overlap between the taste experience of *kokumi* and umami.

konbu - also known as *kombu*; large brown alga (*Saccharina japonica*), an important ingredient in *dashi*. *Konbu* contains large quantities of glutamate and is a source of umami.

kroeung - Cambodian spice mixture used in many Khmer dishes.

lardo - a special type of Italian cold cut made from pork fatback that is cured over a long period of time in marble basins.

lime leaves - also makrut lime and Thai lime (*Citrus hystrix*); citrus tree native to tropical parts of Asia; the leaves and rind are widely used in Thai, Lao, and Cambodian cuisine.

lulas grelhadas - Portuguese expression for grilled squid.

mirin - sweet rice wine with a 14% alcohol content.

miso - Japanese fermented soybean paste.

mouthfeel - see texture.

muscovado sugar - unrefined sugar with a strong molasses taste and high moisture content.

nigiri-sushi - hand-pressed ball of vinegared rice topped with a piece of fish, shellfish, or cephalopod, which in most cases is raw.

nimono - Japanese expression for slow-cooked dishes.

noni - Cambodian name for *Morinda citrifolia*, the leaves of which as used in Asian dishes.

nori - paper thin sheets made from the fronds of the red alga *Porphyra*, which are dried, pressed, and sometimes toasted; among other uses, essential for making sushi rolls.

Pacojet - professional kitchen appliance that consists of a very rapidly rotating, extremely sharp blade that literally shaves a frozen block of food into tiny particles, about five micrometres in size, which is below the threshold where the mouth can detect the individual particles.

pane carasau - traditional twice-baked Sardinian flatbread.

panko - from Japanese *pan* (bread) and *ko* (small pieces); dried, Japanese breadcrumbs that are very light and, therefore, absorb only a little oil when used for deep frying, resulting in a crust that is crisper and less greasy.

Parmigiano Reggiano - the original Italian Parmesan cheese; hard, dry, and aged; has a substantial glutamate content that is a good source of umami.

pescaito frito - Portuguese expression for deep-fried fish, shellfish, and cephalopods.

pesce crudo - Italian expression for raw or nearly raw fish, shellfish, and cephalopods.

piquillo pepper - from the Spanish for little beak; a variety of chili pepper (*Capsicum annuum*) that is sweet and mild.

polbo á feira - (*pulpo a la Gallega*) Galician dish of boiled octopus with paprika and olive oil.

pompia - an unusual large, bright yellow Sardinian citrus fruit (*Citrus limon* var. *pompia*), with a thick, wrinkled skin.

ponzu - a Japanese marinade containing soy sauce, *dashi*, *yuzu* juice, and possibly a little sake.

pulpo a la Gallega - see *polbo á feira.*

pulpo aliñado - Spanish dish of cooked octopus arms in an aliño sauce.

puntarella - an Italian variety of chicory.

puntillitas - (*calamares chiquititos*) Spanish name for the squid species *Alloteuthis subulata*.

rau dang - Vietnamese name for common knotweed, used in soups and hot pots.

saki-ika - dried squid or cuttlefish.

sanbaizu - Japanese marinade made with soy sauce, rice vinegar, and *dashi*.

san-nakji - Korean specialty made with cut-off parts of an arm from an octopus that is alive and that can still move as it is being eaten.

sashimi - Japanese expression for sliced raw fish, shellfish, or cephalopod.

shaoxing - sweet Chinese rice wine used for cooking.

shichimi - Japanese spice blend with seven different tastes; commonly contains *sanshō* pepper, white and black sesame seeds, red chili, ground ginger, *ao-nori* (green seaweed similar to sea lettuce), dried yuzu peel, and hemp seeds.

shiokara - (Japanese for 'salted and spicy') fermentation medium made with the hepatopancreas of squid and cuttlefish.

sorbet - frozen dessert made from fruit juice and sugar, but without dairy products or egg yolks.

sous vide - (French for 'under pressure') term that describes a technique for cooking foods at low temperatures in a tightly sealed plastic pouch.

su - Japanese rice vinegar.

sudako tako - Japanese dish of lightly marinated octopus, for example, with *yuzu* and salt.

surimi - (Japanese for 'minced meat') finely minced meat pressed together into blocks, often made from Humboldt squid (*Dosidicus gigas*) and used as imitation crab and shrimp.

surume - Japanese expression for dried shredded squid or cuttlefish; also a nickname for *Todarodes pacificus*.

tako - Japanese expression for octopus (*Octopus vulgaris*).

tako-yaki - dumplings baked in a special pan made from a flour batter filled with minced, cooked octopus, possibly some *tempura* scraps, pickled ginger, and green onions; served with Japanese mayonnaise and a variety of dipping sauces.

tapas - small cold and hot dishes usually served as snacks with drinks.

tempura - Japanese expression for deep fried fish, shellfish, or vegetables breaded with *panko*.

texture - (mouthfeel) the textural properties of a food that constitute a group of distinctive physical characteristics that can be felt primarily by touch and that are due to the structural elements of the food. Textural properties are connected to mechanical properties such as deformation, breakdown, and streaming of the food when it is subjected to forces in the mouth, such as chewing.

umami - one of five basic tastes. It has two separate aspects, a basal contribution, based on free glutamate, and a strengthening or synergistic contribution, which is due to the presence of 5'-ribonucleotides, including inosinate, adenylate, and guanylate.

ventresca - Italian word for the very fatty tuna belly meat.

wakame - Japanese expression for the brown seaweed *Undaria pinnitifada*.

wasabi - Japanese horseradish (*Wasabi japonica*).

yōkan - Japanese confectionary or candy made from red *azuki* bean paste that is made into a firm gel with the help of sugar and the thickener agar (*kanten*).

yuzu - small Japanese citrus fruit (*Citrus junus*) that has a more aromatic taste than a lemon.

zajin chao xianyou - Cantonese dish of deep-fried squid with sugar peas.

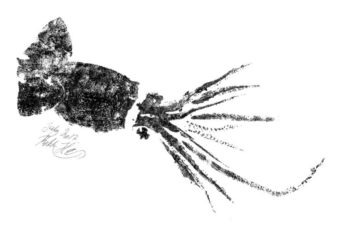

◘ *Gyotaku* of a *Todarodes sagittatus.*

Illustration Credits

Unless noted below, all photos have been taken by Jonas Drotner Mouritsen. Utagawa Kuniyoshi, pp. VIII, 131; Charles Zuckermann, p. XI bottom; Ernst Haeckel, p. 1, 261; depositphotos, p. 5 (Dieter Hawian), p. 54 (Plancton-Video), p. 73 (Valeriy Lebedev), p. 75 (Dmitry Chulov); Heraklion Archaeological Museum p. 6; Smithsonian National Museum of Natural History, p. 8; Steenstrupia 29, 39–47, 2005, p. 8; Pierre Denys de Montfort, p. 10; Tsunemi Kubodera, pp. 11, 90; Katsushika Hokusai, p. 13; Alphonse de Neuville, p. 14; iStock, p. 15 (Ryan J. Lane), p. 17 (Michael Ziegler), p. 39 left (Damocean), p. 39 right (Veliferum), p. 71 (Frank van der Bergh), p. 93 (Nessa-flame); Joshua Lambus, p. 16 ; Xavier Bailly, p. 27; Ole G. Mouritsen, pp. 35, 39 middle, 58, 66, 67, 81, 82, 85, 92, 96, 122 bottom right, 129, 216, 221, 231; Scott Portelli, s. 57; Anders Brix, p. 76; Mathias Porsmose Clausen, p. 106; Adrian Franklin, p. 122 top; Rob Whitrow, p. 122 bottom left; Kristoff Styrbæk, pp. 89, 115, 133, 146, 154, 159, 160, 162, 164, 175, 177; Klavs Styrbæk, pp. 187, 190, 192, 195; Julia Sick, p. 197, 211; Roberto Flore, 204 top; Natural History Museum, London, p. 239; Rikke Højer, p. 260.

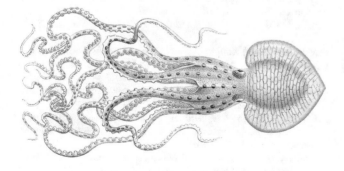

Bibliography

Abollo. E., C. Gestal & S. S. Pascual. Anisakis infestation in marine fish and cephalopods from Galician waters: an updated perspective. *Parasitol. Res.* **87**, 492–499, 2001.

Albertin, C. B., O. S. T. Mitros, Z. Y. Wang, J. R. Pungor, E. Edsinger-Gonzales, S. Brenner, C. W. Ragsdale & D. S. Rokhsar. The octopus genome and the evolution of cephalopod neural and morphological novelties. *Nature* **524**, 220–224, 2015.

Anderson, S. C., J. M. Flemming, R. Watson & H. K. Lotze. Rapid global expansion of invertebrate fisheries: trends, drivers, and ecosystem effects. *PLoS One* **6**, e14735, 2011.

Aristoteles. *Historia animalium* (translated by D. Balme) Cambridge University Press, Cambridge, 2002.

Arkhipkin, A. I. *et al.* World squid fisheries. *Rev. Fish. Sci. Aquac.* **23**, 92-252, 2015.

Borelli, L., F. Gherardi & G. Fiorito. *A Catalogue of Body Patterning in Cephalopoda*. Stazione Zoologica A. Dohrn, Firenze University Press, Firenze, 2005.

Boyle, P. R. & P. G. Rodhouse. *Cephalopods: Ecology and Fisheries.* Blackwell Science, Oxford, 2005.

Bru, R. Tentacles of Love and Death: from Hokusai to Picasso. In *Imágines Secretas: Picasso u la estampa erotica japonese*. Museu Picasso, Barcelona, 2009, p. 194.

Brunt, M. A. & J. E. Davies (eds.) The Cayman Islands Natural History and Biogeography, Springer 1994, p. 175.

Cerullo, M. M. & C. F. E. Roper. *Giant Squid.* Smithsonian, Capstone Press, North Mankato, 2012.

Cousteau, J. & P. Diolé. *Octopus and Squid: The Soft Intelligence.* Doubleday, New York, 1973.

Courage, K. H. *Octopus: The Most Mysterious Creature in the Sea.* Current, New York, 2013.

Cronin, I. *The International Squid Cookbook.* Aris Books, New York, 1981.

Davidson, A. *North Atlantic Seafood.* Prospect Books, Totnes, 2003.

Delgado, D. A. M., M. D. V. Almeida & S. Parisi. *Chemistry of the Mediterranean Diet.* Springer, New York, 2017.

Derby, C. D. Cephalopod ink: production, chemistry, functions, and applications. *Mar. Drugs* **12**, 2700–2730, 2014.

di Cosmo, A. & G. Polese. Neuroendocrine-immune systems response to environmental stressors in the cephalopod *Octopus vulgaris*. *Front. Physiol.* **7**:434, 2016.

Doubleday, Z. A., T. A. A. Prowse, A. Arkhipkin, G. J. Pierce, J. Semmens, M. Steer, S. C. Leporati, S. Lourenço, A. Quetglas, W. Sauer & B. M. Gillanders. Global proliferation of cephalopods. *Cur. Biol.* **26**, R406–R407, 2016.

Ellis, R. *Monsters of the Sea: The History, Natural History, and Mythology of the Oceans' Most Fantastic Creatures*. Doubleday, New York, 1996.

FAO. 2018. *The State of World Fisheries and Aquaculture 2018 – Meeting the sustainable development goals.* Rome. Licence: CC BY-NC-SA 3.0 IGO.

Faxholm, P. L. , C. V. Schmidt, L. B. Brønnum, Y.-T. Sun, M. P. Clausen, R. Flore, K. Olsen & O. G. Mouritsen. Squids of the

North: gastronomy and gastrophysics of Danish squid. *Int. J. Gast. Food. Sci.* **14**, 66–76, 2018.

Fiorito, G. & P. Scotto. Observational learning in *Octopus vulgaris*. *Science* **256**, 545–547, 1992.

Fiorito, G., *et al.* Guidelines for the care and welfare of cephalopods in research – a consensus based on an initiative by CephRes, FELASA, and the Boyd Group. *Lab. Animals* **49**, 1–90, 2015.

García-Fernández, P., M. Prado-Alvarez, M. Nande, D. Garcia de la serrana, C. Perales-Raya, E. Almansa, I. Varó & C. Gestal. Global impact of diet and temperature over aquaculture of *Octopus vulgaris* paralarvae from a transcriptomic approach. *Sci. Rep.* **9**:10312, 2019.

Godfrey-Smith, P. *Other Minds: The Octopus, the Sea, and the Deep Origins of Consciousness.* Farrar, Straus, and Giroux, New York, 2016.

Graça, J., Calheiros & M. M., A. Oliveira. Attached to meat? (Un) Willingness and intentions to adopt a more plant-based diet. *Appetite* **95**, 113–125, 2015.

Grocock, C. & S. Grainger. *Apicius.* Prospect Books, Devon, 2006.

Grossi, G., P. Goglio, A. Vitali & A. G. Williams. Livestock and climate change: impact of livestock on climate and mitigation strategies. *Animal Front.* **9**, 69–76, 2019.

Hachisu, N. S. *Preserving the Japanese Way: Traditions of Salting, Fermenting, and Pickling for the Modern Kitchen.* Andrews McMeel Publishing, LLC, Kansas City, 2015.

Hanlon, R. T. Cephalopod dynamic camouflage. *Curr. Biol.* **17**, R400–R404, 2007.

Hanlon, R. T., R. F. Hixon, J. F. Forsythe & J. P. Hendrix, Jr. Cephalopods attracted to experimental night light during saturation dive at St. Croix. *Bull. Amer. Malocol. Union*, pp. 53–58, 1979.

Hanlon, R. T. & J. B. Messenger. *Cephalopod Behaviour.* Cambridge University Press, Cambridge, 1996.

Heuvelmans, B. *The Kraken and the Colossal Octopus.* Routledge, London, 2006.

Holden-Dye, L., G. Ponte, G. Fiorito, A. L. Allcock, R. Nakajima, E. A. G. Vidal & T. R. Peterson (eds.) *CephsInAction: Towards Future Challenges for Cephalopod Science.* Lausanne: Frontiers Media SA. doi: 10.3389/978-2-88963-437-8, 2020.

Hu., Y., Z. Huang, J. Li & H. Yang. Concentrations of biogenic amines in fish, squid and octopus and their changes during storage. *Food Chem.* **135**, 2604–2611, 2012.

Iglesias, J., F. J. Sánchez, J. G. F. Bersano, J. F. Carrasco, J. Dhont, L. Fuentes, F. Linares, J. L. Muñoz, S. Okumura, J. Roo, T. van der Meeren, E.A.G. Vidal & R. Villanueva. Rearing of *Octopus vulgaris* paralarvae: present status, bottlenecks and trends. *Aquaculture* **266**, 1–15, 2007.

Imperadore, P., S. B. Shah, H. P. Makarenkova & G. Fiorito. Nerve degeneration and regeneration in the cephalopod mollusc *Octopus vulgaris*: the case of the pallial nerve. *Sci. Rep.* **7**:46564, 2017.

Jacquet, J., B. Franks, P. Godfrey-Smith & W. Sánchez-Suárez. The case against octopus farming. *Issues Sci. Technol.* Winter 2019, 37-44, 2019.

Jereb, P. & C. F. E. Roper (eds.) *Cephalopods of the world. An annotated and illustrative catalogue of cephalopod species known to date. Volume 1. Chambered Nautiluses and Sepioids (Nautilidae, Sepiidae,*

Sepiolidae, Sepiadariidae, Idiosepiidae and Spirulidae). FAO Species Catalogue for Fishery Purposes No. 4, Rome, 2005.

Jereb, P. & C. F. E. Roper (eds.) *Cephalopods of the world. An annotated and illustrative catalogue of cephalopod species known to date. Volume 2. Myopsid and Oegopsid Squids.* FAO Species Catalogue for Fishery Purposes No. 4, Rome, 2010.

Jereb, P., C. F. E. Roper, M. D. Norman & J. K. Finn (eds.) *Cephalopods of the world. An annotated and illustrative catalogue of cephalopod species known to date. Volume 3. Octopods and Vampire Squids.* FAO Species Catalogue for Fishery Purposes No. 4, Rome, 2016.

Katsanidis, E. Impact of physical and chemical pretreatments on texture of octopus (*Eledone moschata*). *Food Sci.* **69**, 264–267, 2004.

Katsanidis, E. Physical and chemical pre-treatment of octopuses aiming at tenderisation with reduced processing time and energy requirements. *PTEP* **12**, 45–48, 2008.

Ketnawa, S. & S. Rawdkuen. Application of bromelain extract for muscle food tenderization. *Food Nutr. Sci.* **2**, 393–401, 2011.

Kier, W. M. The musculature of Coleoid cephalopod arms and tentacles. *Front. Cell Dev. Biol.* **4**:10, 2016.

Kier, W. M. & K. K. Smith. Tongues, tentacles and trunks. The biomechanics of movement in molecular hydrostats. *Zoo. J. Linnean Soc.* **83**, 307–324, 1985.

Kier, W. M. & A. M. Smith. The morphology and mechanics of octopus suckers. *Biol. Bull.* **178**, 126–136, 1990.

Kier, W. M. & M. P. Stella. The arrangement and function of octopus arm musculature and connective tissue. *J. Morphol.* **268**, 831–43, 2007.

Kim, M.-K., J.-H. Mah & H.-J. Hwang. Biogenic amine formation and bacterial contribution in fish, squid, and shellfish. *Food Chem.* **116**, 87–95, 2009.

Kröger, B., J. Vinther, & D. Fuchs. Cephalopod origin and evolution. *Bioessays* **33**, 602–613, 2011.

Kubodera, T. & K. Mori. First-ever observations of a live giant squid in the wild. *Proc. Royal Soc. B* **272**, 2583–2586, 2005.

Lee, H. *The Octopus; or, The 'Devil-fish' of Fiction and Fact.* Chapman and Hall, London, 1875.

Levy. G., T. Flash & B. Hochner (2015). Arm coordination in octopus crawling involves unique motor control strategies. *Curr. Biol.* **25**, 1195-1200, 2015.

Mangum, C. P. Respiratory function of the hemocyanins. *Integr. Comp. Biol.* **20**, 19–38, 1980.

Mather, J. A. To boldly go where no mollusc has gone before: personality, play, thinking and consciousness in Cephalopods. *Amer. Maloc. Bull.* **324**, 51–58, 2008.

McFall-Ngai, M. Divining the essence of symbiosis: insights from the squid-vibrio model. *PLoS Biology* **12**, e1001783, 2014.

McGee, H. *On Food and Cooking: The Science and Lore of the Kitchen.* Scribner, New York, 2004.

McGee, H. To cook an octopus: Forget the cork, add science. New York Times, p. F3, 5. marts, 2008.

Middleton, S. *Spineless: Portraits of Marine Invertebrates, the Backbone of Life.* Abrams, New York, 2014.

Mizuta, S., T. Tanaka & R. Yoshinaka. Comparison of collagen types of arm and mantle muscles of the common octopus (*Octopus vulgaris*). *Food Chem.* **81**, 527–532, 2003.

Montgomery, S. *The Octopus Scientists: Exploring the Mind of a Mollusk*. Houghton Mifflin Harcourt, Boston, 2015a.

Montgomery, S. *The Soul of an Octopus: A Surprising Exploration into the Wonder of Conciousness*. Atria Paperback, New York, 2015b.

Morales. J., P. Montero & A. Moral. Isolation and partial characterization of two types of muscle collagen in some cephalopods. *J. Agric. Food Chem.* **48**, 2142–2148, 2000.

Mouritsen, O. G. *Sushi: Food for the Eye, the Body & the Soul*. Springer, New York, 2009.

Mouritsen, O. G. *Seaweeds: Edible, Available & Sustainable*. Chicago University Press, Chicago, 2013.

Mouritsen, O. G. & C. V. Schmidt. A role for macroalgae and cephalopods in sustainable eating. *Front. Psychol.* **11**, 1402, 2020.

Mouritsen, O. G. & K. Styrbæk. *Umami: Unlocking the Secrets of the Fifth Taste*, Columbia University Press, New York, 2014.

Mouritsen, O. G. & K. Styrbæk. *Mouthfeel: How Texture Makes Taste*. Columbia University Press, New York, 2017.

Mouritsen, O. G. & K. Styrbæk. Cephalopod gastronomy — a promise for the future. *Front. Comm. Sci. Environ. Comm.* **3**:38, 2018.

Myhrvold, N. *Modernist Cuisine: The Art and Science of Cooking*. The Cooking Lab Publ., USA, 2010.

O'Doherty Jensen, K. and L. Holm. Preferences, quantities and concerns: socio-cultural perspectives on the gendered consumption of foods. *Eur. J. Clin. Nutr.* **53**, 351–359, 1999.

O'Dor, R. K. & D. M. Webber. The constraints on the celaphods: Why squid aren't fish. *Can. J. Zool.* **64**, 1591–1605, 1986.

Ozogul, Y., O. Duysak, F. Ozogul, A. S. Özkütük & C. Türeli. Seasonal effects in the nutritional quality of the body structural tissue of cephalopods. *Food Chem.* **108**, 847–852, 2008.

Pauly, D., R. Hilborn, & T. A. Branch. Fisheries: Does catch reflect abundance? *Nature* **494**, 303–306, 2013.

Paxton, C. G. M. & R. Holland. Was Steenstrup right? A new interpretation of the 16th century sea monk of the Øresund. *Steenstrupia* **29**, 39–47, 2005.

Payne, A. G., D. J. Agnew & G. J. Pierce. Trends and assessment of cephalopods fisheries. *Fish Res.* **78**, 1–3, 2006.

Prabir, K., M. J. Sarkar & R. Nout (eds.) *Handbook of Indigenous Foods Involving Alkaline Fermentation*. CRC Press, Boca Raton, 2014.

Ramirez, M. D., & T. H. Oakley. Eye-independent, light-activated chromatophore expansion (LACE) and expression of phototransduction genes in the skin of *Octopus bimaculoides*. *J. Exp. Biol.* **218**, 1513–1520, 2015.

Rodhouse, P. G. K., G. J. Pierce, O. C. Nichols, W. H. H. Sauer, A. I. Arkhipkin, V. V. Laptikhovsky, M. R. Lipin'ski, J. E. Ramos, M. Gras, H. Kidokoro, K. Sadayasu, J. Pereira, E. Lefkaditou, C. Pita, M. Gasalla, M. Haimovici, M. Sakai & N. Downey. Environmental effects on cephalopod population dynamics: implications for management of fisheries. *Adv. Mar. Biol.* **67**, 99–233, 2014.

Ruby, M. B. & S. J. Heine. Meat, morals, and masculinity. *Appetite* **56**, 447–450, 2011.

SAPEA, Science Advice for Policy by European Academies. Food from the oceans: how can more food and biomass be obtained from the oceans in a way that does not deprive future generations of their benefits? SAPEA, Berlin, 2017.

Sauer, W. H. H. *et al.* World octopus fisheries. *Rev. Fish. Sci. Aquac.* https://doi.org/10.1080/23308249.2019.1680603, 2019.

Sugiyama, M. *Utilisation of Squid,* A. A. Balkema, Rotterdam, 1989.

Schultz, J. & B. Regardz. *Calamari Cookbook: Exploring the World's Cuisines Through Squid.* Celestialarts, Berkeley, CA, 1987.

Seibel, B. A., S. K. Goffredi, E. V. Thuesen, J. J. Childress & B. H. Robison. Ammonium content and buoyancy in midwater cephalopods. *J. Exp. Marine Biol. Ecol.* **313**, 375–387, 2004.

Scheel, D., P. Godfrey-Smith & M. Lawrence. 2014. *Octopus tetricus* (Mollusca: Cephalopoda) as an ecosystem engineer. *Sci. Mar.* **78**, 521–528, 2014.

Scheel, D., S. Chancellor, M. Hing, M. Lawrence, S. Linquist & P. Godfrey-Smith. A second site occupied by *Octopus tetricus* at high densities, with notes on their ecology and behavior, *Mar. Freshwater Behav. Physiol.* **50**, 285–291, 2017.

Schmidt, C. V. and O. G. Mouritsen. The solution to sustainable eating is not a one-way street. *Front. Psychol.* **11**:531, 2020.

Schmidt, C. V., M. M. Poojary, O. G. Mouritsen, and K. Olsen. Umami potential of Nordic squid (*Loligo forbesii*). *Int. J. Gast. Food Sci* **22**:100275, 2020.

Schmidt, C. V. L. Plankensteiner, P. L. Faxholm, K. Olsen, O. G. Mouritsen & M. B. Frøst. Physicochemical characterisation of sous vide cooked squid *Loligo forbesii* and *Loligo vulgaris*) and the relationship to selected sensory properties and hedonic response. *Int. J. Gast. Food Sci.* **23**:100298, 2021.

Schweid, R. *Octopus.* Reaction Books, London, 2014.

Steenstrup, J. J. S. Om den i Kong Christian IIIs tid i Øresundet fanget Havmund (Sømunken kaldet). *Dansk Maanedsskrift* **1**, 63–96, 1855.

Storelli, M. M., R. Giacominelli-Stuffler, A. Storelli & G. O. Marcotrigiano. Cadmium and mercury in cephalopod molluscs: Estimated weekly intake. *Food Add. Cont.* **23**, 25–30, 2006.

Storelli, M. M., R. Garofalo, D. Giungato & R. Giacominelli-Stuffler. Intake of essential and non-essential elements from consumption of octopus, cuttlefish and squid. *Food Add. Cont. Part B* **3**, 14–18, 2010.

Trueit, T. S. *Octopuses, Squids, and Cuttlefish.* Franklin Watts, New York, 2002.

Tsuji, S. *Japanese Cooking: A Simple Art.* Kodansha International, Tokyo, 1980.

Tye, M. *Tense Bees and Shell-Shocked Crabs: Are Animals Conscious?* Oxford University Press, Oxford, 2016.

Vaz-Pries, P., P. Sixas & A. Barbosa. Aquaculture potential of the common octopus (*Octopus vulgaris* Cuvier, 1797): A review. *Aquaculture* **238**, 221–238, 2004.

Vidal, E. A. G. (ed.) Advances in Cephalopod Science: Biology, Ecology, Cultivation, and Fisheries. *Adv. Mar. Biol.* **76**. Elsevier, Amsterdam, 2014.

Vinther, J., L. Parry, D. E. G. Briggs & P. Van Roy. Ancestral morphology of crown-group molluscs revealed by a new Ordovician stem aculiferan. *Nature* **542**, 471–474, 2017.

Voight, J. R., H. O. Pörtner & R. K. O'Dor. A review of ammonia-mediated buoyancy in squids (Cephalopoda: Teuthoidea). *Mar. Freshwater Behav. Physiol.* **25**, 193–203, 1995.

Wells, M. J. *Octopus: Physiology and Behaviour of an Advanced Invertebrate*. Chapman and Hall, London, 1978.

Williams, W. *Kraken: The Courious, Exciting and Slightly Disturbing Science of Squid*. Abrahams Image, New York, 2010.

Winkelmann, I. F., P. F. Campos, J. Strugnell, Y. Cherel, P. J. Smith, T. Kubodera, L. Allcock, M.-L. Kampmann, H. Schroeder, A. Guerra, M. Norman, J. Finn, D. Ingrao, M. Clarke & M. T. P. Gilbert. Mitochondrial genome diversity and population structure of the giant squid *Architeuthis*: genetics sheds new light on one of the most enigmatic marine species. *Proc. Royal Soc. B* **280**, 20130273, 2013.

World Congress on Cephalopods: Overview on Supplies. Vigo, Spain, 2016.

Young, J. Z. *The Anatomy of the Nervous System of 'Octopus vulgaris.'* Oxford University Press, Oxford, 1971.

Young, R. E., M. Vecchione & D. T. Donovan, The evolution of coleoid cephalopods and their present biodiversity and ecology. *South African J. Mar. Sci.* **20**, 393–420, 1998.

Willett, W. *et al.* Food in the Anthropocene: the EAT–Lancet Commission on healthy diets from sustainable food systems. *Lancet* **393**, 447–492, 2019.

Wrangham, R. *Catching Fire: How Cooking Made Us Human*. Basic Books, New York, 2009.

Index

Printed in the United States
by Baker & Taylor Publisher Services